Lecture Notes in Computer Science 12996

More information about this subseries at https://link.springer.com/bookseries/7408

Mohamed Adel Serhani · Liang-Jie Zhang (Eds.)

Services – SERVICES 2021

17th World Congress
Held as Part of the Services Conference Federation, SCF 2021
Virtual Event, December 10–14, 2021
Proceedings

 Springer

Editors
Mohamed Adel Serhani ⓘ
United Arab Emirates University
Al Ain, United Arab Emirates

Liang-Jie Zhang ⓘ
Kingdee International Software Group
Co., Ltd.
Shenzhen, China

ISSN 0302-9743   ISSN 1611-3349 (electronic)
Lecture Notes in Computer Science
ISBN 978-3-030-96584-6   ISBN 978-3-030-96585-3 (eBook)
https://doi.org/10.1007/978-3-030-96585-3

LNCS Sublibrary: SL2 – Programming and Software Engineering

This Springer imprint is published by the registered company Springer Nature Switzerland AG
The registered company address is: Gewerbestrasse 11, 6330 Cham, Switzerland

Preface

The World Congress on Services (SERVICES) aims to provide an international forum to attract researchers, practitioners, and industry business leaders in all services sectors to help define and shape the modernization strategy and directions of the services industry.

SERVICES is a member of Services Conference Federation (SCF). SCF 2021 featured the following 10 collocated service-oriented sister conferences: the International Conference on Web Services (ICWS 2021), the International Conference on Cloud Computing (CLOUD 2021), the International Conference on Services Computing (SCC 2021), the International Conference on Big Data (BigData 2021), the International Conference on AI and Mobile Services (AIMS 2021), the World Congress on Services (SERVICES 2021), the International Conference on Internet of Things (ICIOT 2021), the International Conference on Cognitive Computing (ICCC 2021), the International Conference on Edge Computing (EDGE 2021), and the International Conference on Blockchain (ICBC 2021).

This volume presents the accepted papers for SERVICES 2021, held as a fully virtual conference during December 10–14, 2021. The major topics of SERVICES 2021 included, but were not limited to, advertising services, banking services, broadcasting and cable TV service, business services, communications services, government services, real estate operations services, schools and education services, healthcare services, etc.

We accepted nine papers, comprising seven full papers and two short papers. Each paper was reviewed by at least three independent members of the SERVICES 2021 international Program Committee. We are pleased to thank the authors whose submissions and participation made this conference possible. We also want to express our thanks to the Organizing Committee and Program Committee members for their dedication in helping to organize the conference and reviewing the submissions. We owe special thanks to the keynote speakers for their impressive speeches.

December 2021

Mohamed Adel Serhani
J. Leon Zhao
Liang-Jie Zhang

Organization

General Chair

J. Leon Zhao   Chinese University of Hong Kong, China

Program Chair

Mohamed Adel Serhani   United Arab Emirates University, UAE

Services Conference Federation (SCF 2021)

General Chairs

Wu Chou   Essenlix Corporation, USA
Calton Pu (Co-chair)   Georgia Tech, USA
Dimitrios Georgakopoulos   Swinburne University of Technology, Australia

Program Chairs

Liang-Jie Zhang   Kingdee International Software Group Co., Ltd.,
                                        China
Ali Arsanjani   Amazon Web Services, USA

Industry Track Chairs

Awel Dico   Etihad Airways, UAE
Rajesh Subramanyan   Amazon Web Services, USA
Siva Kantamneni   Deloitte Consulting, USA

CFO

Min Luo   Georgia Tech, USA

Industry Exhibit and International Affairs Chair

Zhixiong Chen   Mercy College, USA

Operations Committee

Jing Zeng	China Gridcom Co., Ltd., China
Yishuang Ning	Tsinghua University, China
Sheng He	Tsinghua University, China

Steering Committee

Calton Pu (Co-chair)	Georgia Tech, USA
Liang-Jie Zhang (Co-chair)	Kingdee International Software Group Co., Ltd., China

SERVICES 2021 Program Committee

Changyu Dou	Guangdong Polytechnic Normal University, China
Xiaohu Fan	Wuhan Collage, China
Qin Guo	Shenzhen University, China
Jinhong Xu	Shenzhen Institute of Information Technology, China
Hasan Ali Khattak	National University of Sciences and Technology, Pakistan
Chao Li	Beijing Jiaotong University, China
Xiuhua Li	Chongqing University, China
Xin Luo	Chongqing Institute of Green and Intelligent Technology, China
Hu Lv	Shenzhen Polytechnic, China
Yiguan Ma	Shenzhen Polytechnic, China
Waseem Mufti	Aalborg University, Denmark
Feng Qiu	University of Shenzhen, China
Sérgio Ribeiro	CPQD, Brazil
Stefano Sebastio	Raytheon Technologies, USA
Vikas Shah	Knights of Columbus, USA
Yifei Wu	Shenzhen Polytechnic, China
Zhu Xiangbo	Shenzhen Polytechnic, China
Li Yi	Shenzhen Institute of Information Technology, China
Kunjing Zhang	Institute of Information and Technology, China
Lindong Zhao	Shenzhen Municipal Human Resources and Social Security Bureau, China

Conference Sponsor – Services Society

The Services Society (S2) is a non-profit professional organization that has been created to promote worldwide research and technical collaboration in services innovations among academia and industrial professionals. Its members are volunteers from industry and academia with common interests. S2 is registered in the USA as a "501(c) organization", which means that it is an American tax-exempt non-profit organization. S2 collaborates with other professional organizations to sponsor or co-sponsor conferences and to promote an effective services curriculum in colleges and universities. S2 initiates and promotes a "Services University" program worldwide to bridge the gap between industrial needs and university instruction.

The services sector accounted for 79.5% of the GDP of the USA in 2016. Hong Kong has one of the world's most service-oriented economies, with the services sector accounting for more than 90% of GDP. As such, the Services Society has formed 10 Special Interest Groups (SIGs) to support technology and domain specific professional activities:

- Special Interest Group on Web Services (SIG-WS)
- Special Interest Group on Services Computing (SIG-SC)
- Special Interest Group on Services Industry (SIG-SI)
- Special Interest Group on Big Data (SIG-BD)
- Special Interest Group on Cloud Computing (SIG-CLOUD)
- Special Interest Group on Artificial Intelligence (SIG-AI)
- Special Interest Group on Edge Computing (SIG-EC)
- Special Interest Group on Cognitive Computing (SIG-CC)
- Special Interest Group on Blockchain (SIG-BC)
- Special Interest Group on Internet of Things (SIG-IOT)

About the Services Conference Federation (SCF)

As the founding member of the Services Conference Federation (SCF), the First International Conference on Web Services (ICWS) was held in June 2003 in Las Vegas, USA. A sister event, the First International Conference on Web Services - Europe 2003 (ICWS-Europe 2003), was held in Germany in October of the same year. In 2004, ICWS-Europe was changed to the European Conference on Web Services (ECOWS), which was held in Erfurt, Germany. The 19th edition in the conference series, SCF 2021, was held virtually over the Internet during December 10–14, 2021.

In the past 18 years, the ICWS community has expanded from Web engineering innovations to scientific research for the whole services industry. The service delivery platforms have expanded to mobile platforms, the Internet of Things (IoT), cloud computing, and edge computing. The services ecosystem has gradually been enabled, value added, and intelligence embedded through enabling technologies such as big data, artificial intelligence, and cognitive computing. In the coming years, transactions with multiple parties involved will be transformed by blockchain.

Based on the technology trends and best practices in the field, SCF will continue serving as the conference umbrella's code name for all services-related conferences. SCF 2021 defined the future of the New ABCDE (AI, Blockchain, Cloud, big Data, Everything is connected), which enable IoT and support the "5G for Services Era". SCF 2021 featured 10 collocated conferences all centered around the topic of "services", each focusing on exploring different themes (e.g. web-based services, cloud-based services, big data-based services, services innovation lifecycle, AI-driven ubiquitous services, blockchain-driven trust service-ecosystems, industry-specific services and applications, and emerging service-oriented technologies). The SCF 2021 members were as follows:

1. The 2021 International Conference on Web Services (ICWS 2021, http://icws.org/), which was the flagship conference for web-based services featuring web services modeling, development, publishing, discovery, composition, testing, adaptation, and delivery, as well as the latest API standards.
2. The 2021 International Conference on Cloud Computing (CLOUD 2021, http://the cloudcomputing.org/), which was the flagship conference for modeling, developing, publishing, monitoring, managing, and delivering XaaS (everything as a service) in the context of various types of cloud environments.
3. The 2021 International Conference on Big Data (BigData 2021, http://bigdataco ngress.org/), which focused on the scientific and engineering innovations of big data.
4. The 2021 International Conference on Services Computing (SCC 2021, http://the scc.org/), which was the flagship conference for the services innovation lifecycle including enterprise modeling, business consulting, solution creation, services orchestration, services optimization, services management, services marketing, and business process integration and management.

5. The 2021 International Conference on AI and Mobile Services (AIMS 2021, http://ai1000.org/), which addressed the science and technology of artificial intelligence and the development, publication, discovery, orchestration, invocation, testing, delivery, and certification of AI-enabled services and mobile applications.
6. The 2021 World Congress on Services (SERVICES 2021, http://servicescongress.org/), which put its focus on emerging service-oriented technologies and industry-specific services and solutions.
7. The 2021 International Conference on Cognitive Computing (ICCC 2021, http://thecognitivecomputing.org/), which put its focus on Sensing Intelligence (SI) as a Service (SIaaS), making a system listen, speak, see, smell, taste, understand, interact, and/or walk, in the context of scientific research and engineering solutions.
8. The 2021 International Conference on Internet of Things (ICIOT 2021, http://iciot.org/), which addressed the creation of IoT technologies and the development of IOT services.
9. The 2021 International Conference on Edge Computing (EDGE 2021, http://theedgecomputing.org/), which put its focus on the state of the art and practice of edge computing including, but not limited to, localized resource sharing, connections with the cloud, and 5G devices and applications.
10. The 2021 International Conference on Blockchain (ICBC 2021, http://blockchain1000.org/), which concentrated on blockchain-based services and enabling technologies.

Some of the highlights of SCF 2021 were as follows:

- Bigger Platform: The 10 collocated conferences (SCF 2021) got sponsorship from the Services Society which is the world-leading not-for-profits organization (501 c(3)) dedicated to serving more than 30,000 services computing researchers and practitioners worldwide. A bigger platform means bigger opportunities for all volunteers, authors, and participants. In addition, Springer provided sponsorship for best paper awards and other professional activities. All 10 conference proceedings of SCF 2021 will be published by Springer and indexed in the ISI Conference Proceedings Citation Index (included in Web of Science), the Engineering Index EI (Compendex and Inspec databases), DBLP, Google Scholar, IO-Port, MathSciNet, Scopus, and ZBlMath.
- Brighter Future: While celebrating the 2021 version of ICWS, SCF 2021 highlighted the Fourth International Conference on Blockchain (ICBC 2021) to build the fundamental infrastructure for enabling secure and trusted services ecosystems. It will also lead our community members to create their own brighter future.
- Better Model: SCF 2021 continued to leverage the invented Conference Blockchain Model (CBM) to innovate the organizing practices for all 10 collocated conferences.

Contents

Communication with Non-native Speakers Through the Service of Speech-To-Speech Interpreting Systems: Testing Technology Capacity and Exploring Specialists' Views

Anastasia Atabekova (✉) (ORCID)

Peoples' Friendship University of Russia, Moscow 117198, Russia
atabekova-aa@rudn.ru

Abstract. The paper explores speech-to-speech interpretation systems (SISs) application to service multilingual communication in humanitarian contexts related to forced migration. The study compares various systems capacity to process non-native speakers' speech to map major challenges for the above systems use within the mentioned settings. The research introduces interim results of a pilot study in terms of research sample (non-native speakers and interpreters), the selected language pair, and the comparative list of recommender platforms. The research includes the relevant literature study, comparative analysis of SISs outputs regarding the interpretation of non-native speakers' accented speech, when interpreted from English into Russian, interpreters' surveys on the above analysis results. The technology under study included Google Translator, Microsoft Translator, and Yandex. The pool of research participants included refugees from different countries and professional interpreters. The research rested on comparative qualitative multidimensional analysis, integrated content-based selection of academic sources and their theoretical analysis, descriptive empirical analysis of language errors by SISs, interpreters' survey through open-ended questionnaire, factor, cluster, and content analysis to process their replies. The results map those language and communicative context features that should be considered for digital interpreting systems further tuning in terms of multilingual instrumentation, set forth the tasks to develop the relevant methodology for further studies, customize the technology to specific communicative settings and to train specialists to use it in socially critical contexts.

Keywords: Multilingual communication service · Speech-to speech interpretation systems · Non-native speakers

1 Introduction

The research on SISs stands within the broader studies on artificial intelligence (further AI) applications that constantly expand their operation within industrial, social, governmental areas, etc. [3]. The recent developments advance AI tools to perceive, generate, translate, and interpret human speech. Meanwhile, the 21st century faces peoples' intensive migration amid emergences, including natural disasters and man-made disorders.

© Springer Nature Switzerland AG 2022
M. A. Serhani and L.-J. Zhang (Eds.): SERVICES 2021, LNCS 12996, pp. 1–17, 2022.
https://doi.org/10.1007/978-3-030-96585-3_1

Such a tense landscape puts on the agenda the need for digital solutions to serve communication in respective emergency contexts, which include interaction among foreigners (who even might turn out to be undocumented forced migrants) and hosting country representatives in healthcare, administrative, legal, educational, social settings. Due to possible emergency, respective interaction lack due language support, meaningful information, etc. [16]. Therefore, the actors use open access and free SISs that spread faster than AI gurus tailor them to specific needs. The above state of affairs often leads to the use of English by non-native speakers in the working language pair where the second one is the hosting country language, which local officials use. This situation raises the question of SISs quality and fidelity when the source text is produced by non-native speakers. The above explains the research topicality and requires consideration of current trends in the respective areas of research and industry.

Modern research lays special emphasis on AI applications in natural speech recognition [27], acoustic-phonetic and phonological interpretation [1], statistical approach to study speech recognition [9], the interrelation of AI tools and cognitive semantics [13], natural language processing with AI use [2]. Particular emphasis is laid on speech interpretation systems (further SIS) applications, including SIS use for the conversion of a speaker's utterance into a text by a web browser [31], the analysis of semantic interpretations produced by humans and SISs [14], the study of tailored (commercial) and open access systems [15], the outcomes of the use of intelligent personal assistants for native and non-native language speakers [34], racial disparities with reference to SISs use [17]. Comparative analysis of different virtual assistants has also resulted in pathways for respective tools improvement, regarding Speech to Text (STT) and Text to Speech (TTS) technologies [23]. Contemporary studies explore the accuracy rate for various languages translations provided by speech-to-text recognition (STR) and computer-aided translation (CAT) systems [2]. In terms of methodology, a specific attention is paid to standard learning process for randomized neural networks [25]. The above trends reveal that that currently applied research moves on without systemic attention to languages and specific features of communication settings. However, experts underline the importance of further stance on customizing language models that are specific to a particular domain, environment, and accents [34]. This agenda also addresses digital support for interpreting and translation within crisis management issues from the angle of human rights provision [22]. However, neither academic papers have been found regarding the interpreting capacity of digital tools regarding their use to service non-native speakers' communicative needs in emergency humanitarian contexts. Nor interpreters' voices from the ground have been explored. Meanwhile, such trends seem timely due to increased migration within global scale. The above landscape confirms a current importance and social relevance of the theme under consideration and justifies the format of pilot study that is expected to lay preliminary grounds for further systemic analysis. The results of literature review have shaped the research goals, questions, and pathway for empirical studies.

The research goal is two-fold and aims to compare speech interpretation systems in terms of their fidelity when processing non-native speakers' speech in humanitarian contexts and to explore language service providers' experience in relation to the respective theme. The above objective requires consideration of the following research questions:

RQ1. What are the capacities of speech-to speech systems for interpreting non-native speakers' accented speech from English into Russian?

RQ2. What are interpreters' opinions of the above system capacities and what features they consider necessary to consider for the speech interpretation technology enhancement, and training of specialists who are to use the mentioned tools?

The replies to the above questions are to lay grounds for preliminary conclusions on the relevant methodology development and recommendations for specialists who deal use the technology under study for professional purposes.

2 Research Stages and Methodology

The research was conducted during 2019–2020. In line with the research questions the study integrated qualitative and quantitative techniques, included several stages, that focused on different actors, used different data, procedures, and methods.

2.1 Stage One

The experiment aimed to reply to the first research question on the results of systems use for interpreting non-native speakers' accented speech from English into Russian.

The choice of language pair is due to a few reasons. First, these languages belong to the official UNO languages, and are use within international operation in humanitarian contexts. Second, the author feels competent to work with the languages due to her professional certification in English, and status of Russian native speaker. *The technology under study* included Google (https://cloud.google.com/speech-to-text/docs/languages), Microsoft (https://www.microsoft.com/en-us/translator/languages.aspx), and Yandex tools (https://yandex.ru/support/translate/supported-langs.html). They have real-time SIS apps and most widely used due to free open access.

Data collection included the video recordings of speeches of non-native speakers of English. The videos were taken from the YouTube open access platform. The materials were selected with regard to the population of forced migrants who moved to other countries due to armed conflicts and/or socio-political unrest. Initially 39 videos with 81 English speeches of refugees were found. However, even the preliminary check of the speech interpretation systems capacity resulted in the cut of this research sample by one third due to the amount of speech items that the systems under study failed to interpret. Finally, 11 videos with 22 non-native refugee speakers were subject to study. Transcripts for the texts of all the speakers were made regarding the syntagmatic division of the speech that operates by the concept of syntagm [24, p. 86] which means a linear sequence or pattern of linguistic objects that are contextually united by horizontal relations in terms of semantics, syntax, morphology, phonetics, and intonation [32]. Totally 974 syntagms were considered in terms of their interpretation by three SIS tools. The number and type of errors (lexical, grammatical) in each variant of interpretation was analyzed. The number of syntagms was considered sufficient for the pilot study [10]. The Appendix lists the resources, the length of speeches and the number of syntagms in each analyzed speech slot.

Research Sample Variables. The speakers' variables covered the following features: country origin, sex, approximate age boundaries, speech features with regard to communicative contexts, event settings, emotional state, accent. The videos include representatives of different African regions and countries, Afghanistan, Pakistan, Syria, North Korea, Myanmar. There are four female and 18 male speakers. The population age varies from adolescents (until 19), young people (until the age of 24), according to the World Health Organization criteria (https://www.who.int/southeastasia/health-topics/adolescent-health), and adults. As for the refugees' emotional state, the following physically justified features were identified by the author and further checked up for unified classification with the interpreters who took part in the survey at the next stage of the study: quite coherent speech/broken separate phrases and/incoherent speech/tears/words and/or facial expressions of sadness/disappointment/suffering and misery.

The study took into the account the point that the dependence of degree of accent on various communicative contexts [21]. The present research does not aim to explore the accent with respect to SISs fidelity and considers speech variables regarding communicative contexts which include structured video report/scheduled interview/preliminary awareness of the questions content and/or procedure/spontaneous replies to the journalist's questions. The event settings variables cover illegal border crossing, refugees' move in the border crossing area, sea rescue operations, refugee camps every-day routine, pre-pared video message, public speech in solemn ceremony settings.

Data analysis aimed to close the experiment to real world settings. A loudspeaker, connected to the computer was used as a source speaking English; the speech was transmitted through the loudspeaker that was separated from the tablet for 3 cm. The procedure of the experiment included the testing of the samples by the three mentioned systems. After each speech extract interpretation its transcript was copied from the screen and fixed in written form. The scope of the paper did not allow produce the visual results of all samples investigation. The analysis of one of them is considered in the next section as an example. The experiment used BLUE [7] and WER [2] methods to evaluate the quality of speech samples generated by SIS under study. The level of correctness was evaluated on grounds of the lexical and the grammatical accuracy within the syntagm in the target language utterance. The quality of the accuracy in the comparison was determined by statistical calculation.

2.2 Stage Two

The analysis included a pilot survey of interpreters in line with the second research question.

Research sample included the participants to the survey who were professional interpreters invited by e-mail messages to discuss the SISs outputs. Initially over 200 interpreters expressed their interest. The selection criteria included affiliation with the Russian language native community, experience in public service interpreting for foreigners, length of employment, interpreting experience in context related to healthcare, administrative, legal contexts that go beyond social standards of communication, interpreting awareness of and experience in SISs use. Finally, 56 interpreters were invited to take part in the pilot survey. The number of respondents was considered sufficient for a pilot study under the existing criteria [10]. All the respondents had university degree in translation and interpreting, over 5-year working experience. Their working languages include Arabic, Chinese, French, German, Dari, Pashtu, English, Uzbek, Tajik. All those engaged confirmed their working experience in instant interpreting for public services without previous arrangements with other actors, interpreting in crisis situations related to natural disasters, armed conflicts, social and political turmoil.

Data collection included several steps. First, the respondents were made familiar with videos, were asked to produce their comments on speakers' variables to check up and precise the initial data specified by the author. The data concerned mostly the degree of speakers' accent and the verbal description of their emotional state. Further the respondents were informed about the results of the systems use for refugees' speech interpretation and were invited to reply to the following open-end questions:

1. Do you use AI-based speech interpretation systems (SISs)in your work? (yes/no, please, specify reasons and situations).
2. Did your experience in public service interpreting include emergency work with migrants? (yes/no, please, specify contexts and situations).
3. What were target audiences for who you mediated?
4. What tools did you use? (please, specify reasons for your choice).
5. Were you satisfied with the mentioned tools? (please, specify why).
6. Why do you think of the presented experiment results? (free style comments).
7. What should be considered to design and improve SISs bearing in mind the present research findings? (please, provide free style comments regarding the point to be considered)
8. What recommendations can be shaped to for specialists who deal with SISs?

Data analysis was conducted in line with to the second research question and used open-ended questionnaire to get background cluster and factor analysis of replies helped to structure respondents' experience in working with SISs in humanitarian emergency contexts. Conventional and Summative Content analysis of free-style written comments was implemented to identify key concepts and the level of their relevance regarding interpreters' vision on reasons, contexts, and limitations for SIS. Statistical Package for the Social Sciences (SPSS) was used to process statistic data.

3   Results and Discussion

3.1   Capacities of Speech-to Speech Systems for Interpreting Non-native Speakers' Accented Speech from English into Russian

The experiment made it possible to analyze the procedure of speech recognition systems work, their features and operation specifics regarding different actors and settings. The common characteristics of such speech interpretation systems as Google Translator, Microsoft Translator Speech API and Yandex Speech Interpretation platform are the following. The SISs do not recognize the flow of speech immediately as the audio file for speech recognition is sent to the cloud service first. Moreover, the interpreting platforms face challenges to identify utterances produced with accent. However, the current findings showed that the Google Translator and Microsoft Translator show more capacity to interpret the utterance in a foreign language and make a generated speech sample in the target language automatically.

Table 1 shows how the Google, Microsoft, and Yandex recognize the source audio texts in English and transform them into Russian. The interpreted segments have certain indices in the tables.

In many of the above syntagms, Syrian accent was not recognized at all i.e., no fidelity occurred. Many difficulties also arose in connection with geographical names, the latter were either not recognized or were recognized as other nouns. Microsoft recognized or "tried" to recognize all semantical segments though complete fidelity was achieved in two segments. Google Translator could not recognize the meaning of six segments. A complete fidelity was achieved only once. In all other cases, partial fidelity was achieved. All the video reports were subject to transcript of speech interpretation results in the above way. Another criterion of speech generation was the quantity of lexical and grammar mistakes made in a generated sample. If a lexical unit in a target speech did not match the lexical unit in the source speech it was considered as a mistake. Grammar mistakes included not logically justified changes of the parts of speech and syntactical word order. If a speech segment was not detected, all lexical units that comprise the segment were assumed as mistakes and grammar mistakes were identified as 2 points. The same approach was used when the speech segment was detected but no fidelity was achieved. Finally, the average data were calculated to compare the SISs.

The average data showed that SISs make grammar mistakes in almost all syntagms. The number of lexical mistakes made by Yandex Interpretation system also exceeds the results of other systems. The least number of mistakes was made by Microsoft Translator Speech API, namely 2,72 lexical mistakes and 0, 9 grammar mistakes for 11 syntagms on average.

The data regarding 11 videos with 22 speakers, their features, and errors in interpretation of their speech per syntagma from English into Russian by three tools goes bellow in Table 3 and provides results that are compatible with the above characteristics regarding the percentage of errors made by the SISs under study.

Three videos in the research sample outstand by the number of errors committed by all the SISs. Thus, video 5 refers to the sea rescue operation, shows the emergency on a rescue boat, where female and male speakers from African countries in emergency, experiencing post-survival shock with little belief in future steps toward safe life, while having just been saved from the downing boat at sea. Their speech in English is produced with a strong accent. Video 10 portrays a female adolescent who in tears delivers a prepared speech at a high level international public ceremony and talks about her poor experience and reasons to flee from home country. Her speech is characterized by an obvious accent. Video 11 is a kind of video interview which delivers the prepared monologue by a person with a very strong accent. The respective figures produce the evidence on the features that affect SISs output quality. The data from 11 videos and 22 speakers' productions reveals that such specifics as non-English native speakers' origin, sex, age, mother tongue do not produce critical impact on the speech interpretation systems capacity. The present findings show that SIS output quality depends primarily on the speaker's his/her emotional state, degree of his/her speech preparedness, social settings of communicative context. In other words, the current error -based analysis reveals that the physical and social settings are more

Table 1. The output of speech interpretation (SI) from English into Russian by three SIS of a Syrian refugee who speaks in the border crossing area (video 1), equivalent in English goes first for the readers' convenience, Author's data.

Original syntagms	Microsoft SI into Russian	Google SI into Russian	Yandex SI into Russian
1 The road for three years	[I don't know yet] Я ещё не знаю	*No text generated*	*No text generated*
2 in Jordan and then to Turkey	[Yeah, than Jordan, and then] Да, Джордан, а потом	[Jordan] Иордания	[Jordan] Иордания
3 and then Greece, Makedonia and then here	[And then agrees on Magnolia and then here] [А потом соглашается на Магнолия, а затем здесь]	[Anthony makris] Энтони макрис	[Other than that, Macedonia and the] Кроме этого, видео, Македония
4 And why don't you stay here in Turkey?	[Why did you stay in Turkey?] Почему вы остаетесь в Турции?	[why don't you stay on Tekkit?] Почему бы вам не остаться на Текките?	[So] Так
5 Because you cannot work or	[be cause you cannot work or] быть причиной вы не можете работать или	*No text generated*	*No text generated*
6 study or go out	[study or] учиться или	*No text generated*	*No text generated*
7 And Germany or Britain or Sweden	[Germany orbit north Sweden] И Германия орбита Северная Швеция	[Germany on] Германия на	[They were ready] Они были готовы
8 sends you back again?	[they send you back again]	*No text generated*	*No text generated*
9 I will try,	[I will try] Я попытаюсь	[I will try] Я попробую	[There are 2 and 1] Есть 2 и 1
10 We will try again	[We will try] Мы попытаемся	*No text generated*	[Credited] Кредитованный
11 All Syrian people now live on the face of God	[All of the city and people now live...] Весь город и люди теперь живут	*No text generated*	*No text generated*

Table 2. Errors made in the interpretation into target language (Russian), Author's data

Syntagm N	Microsoft SIS		Google SIS		Yandex SIS	
	Lexical	Grammar	Lexical	Grammar	Lexical	Grammar
1	5	2	5	2	5	2
2	3	1	3	2	4	2
3	3	1	5	2	5	2
4	3	1	4	0	5	1
5	1	1	3	2	4	2
6	2	0	4	2	4	2
7	4	1	4	1	5	1
8	1	1	3	2	4	1
9	0	0	0	0	2	0
10	1	1	2	2	1	2
11	7	0	4	2	6	2
Errors	2,72	0,9	3,36	1,55	4,09	1,54

critical that the type of accent which practically does not matter. The SISs produce better results within the following settings:

- speech production by non-native speakers out of emergency settings,
- spontaneous replies without visually exposed emotional tension (shock, anxiety),
- spontaneous replies or comments in the context of non-native speakers' awareness of the procedure,
- spontaneous replies or comments by non-native speakers' who have some ideas, shaped in advance of the communication process,
- speech production by non-native speakers prepared in advance and delivered to a camera person.

The results of speech extracts recognition (see Tables 1, 2 and 3) confirm that it is not just the question of language units' recognition but the challenge the systems face regarding need for mapping a particular situation context, its actors, and constitutive concepts that they mean when use language units. This problem has been mentioned in earlier research. For instance, scholars underline that with regard to work with refugees, the design of chatbots should consider refugees' needs [4]. Moreover, it seems relevant to mention the data obtained by scholars who considered the SIS use in 911 emergency service [28], and the application of the digital tools for speech interpretation within medical settings [26]. The present research adds new empirical evidence regarding concrete forced migration emergency settings, thus enhancing the data on the investigation of the context-dependent speech recognition and fostering the idea of the need to further develop automated adaptive speech tutoring, that has been specified on general grounds

Table 3. Average number of errors per syntagma in interpretation of each speaker's production, Author's data

N of video and speaker	Speaker's features (if mentioned), communication context event settings		Average number of errors per syntagma					
			Microsoft SIS		Google SIS		Yandex SIS	
			Lex	*Gr*	*Lex*	*Gr*	*Lex*	*Gr*

Video 1. Refugees' move in the border crossing area, no visual emotional tension, spontaneous replies within the awareness of the questioning procedure, slight and obvious accents

N of video and speaker	Speaker's features		Microsoft Lex	Microsoft Gr	Google Lex	Google Gr	Yandex Lex	Yandex Gr
Speaker 1	Syria	Young man	2,72	0,9	3,36	1,55	4,09	1,54
Speaker 2	Syria	Young man	2,5	1,2	3,1	1,36	3,9	1,4

Video 2. Illegal Border Crossing, spontaneous replies, no visual emotional tension, obvious accents

Speaker 1	Pakistan	Adult man	2,43	1,24	3,2	1,4	3,8	1,7
Speaker 2	Pakistan	Adult man	2,34	1,23	3,05	1,3	3,9	1,5

Video 3. Sea Rescue Operation, emergency situation (refugees just taken from downing boats), visual emotional tension, anxiety, spontaneous replies, strong accents

Speaker 1	African continent	Young man	4,4	1,5	5,6	2,49	5,9	2,7
Speaker 2	African continent	Adult woman	4,7	1.7	5,9	2,53	6,2	2,6
Speaker 3	African continent	Young woman	4,5	1,9	5,8	2,57	6,4	2,9

Video 4. Sea Rescue Operation, safety after emergency, no visual emotional tension, awareness of the interview procedure, quiet coherent speech, sadness, slight accent

Speaker 1	Gambia	Young man	2,6	1,2	3,3	1,3	3,7	1,5

Video 5. Refugee Camp everyday routine, no visual emotional tension, spontaneous replies, disappointment, obvious accents

Speaker 1	Syria	Adult man	2,7	1,4	3,4	1,4	3,7	1,8
Speaker 2	Syria	Male adolescent	2,6	1,29	3,4	1,5	3,8	1,6
Speaker 3	Syria	Young man	2,7	1,22	3,5	1,4	3,73	1,7
Speaker 4	Afghanistan	Young man	2,8	1,18	3,3	1,5	3,6	1,7
Speaker 5	Pakistan	Adult man	2,9	1,32	3,4	1,4	3,7	1,7

Video 6. Refugee Camp everyday routine, no visual emotional tension, spontaneous replies within the awareness of the questioning procedure, slight and obvious accents

Speaker 1	Afghanistan	Male adolescent	2,4	1,32	3,8	1,5	4,2	1,5
Speaker 2	Afghanistan	Female adolescent	2,6	1,24	3,4	1,3	3,7	1,4

(continued)

Table 3. (*continued*)

N of video and speaker	Speaker's features (if mentioned), communication context event settings		Average number of errors per syntagma					
			Microsoft SIS		Google SIS		Yandex SIS	
			Lex	*Gr*	*Lex*	*Gr*	*Lex*	*Gr*
Video 7. Refugee Camp everyday routine, prepared comments, no visual emotional tension, disappointment, slight and obvious accents								
Speaker 1	Gambia	Young (27) man	2,4	1,21	3,3	1,51	3,4	1,8
Speaker 2	Gambia	Adult (9 over 30+) man	2,8	1,3	3,6	1,6	3,8	1,7
Video 8. Refugee Camp for Minors, official structured video, prepared speech, no visual emotional tension, sadness, slight accent								
Speaker 1	Afghanistan	Male adolescent	2,5	1,2	3,4	1,5	3,8	1,9
Speaker 2	Syria	Male adolescent	2,4	1,3	3,3	1,6	3,9	1,8
Video 9. Administrative Settings, awareness of the questioning procedure, no visual emotional tension, obvious accent								
Speaker 1	Rohingya	Male boy/adolescent	2,8	1,25	3,4	1,5	4,1	1,9
Video 10. Solemn Public Ceremony Settings, prepared speech, emotional tension, tears, partly broken speech, obvious accent								
Speaker 1	North Korea	Female adolescent	5,5	1,5	6,0	2,1	7,7	2,7
Video 11. Video message, structured video, prepared speech, no visual emotional tension, sadness, strong accent								
Speaker 1	Myanmar's Chin people refugees' representative	Adult man	3,2	1,3	4,4	1,4	4,8	1,8
Average number of errors			3,02	1,33	3,86	1,62	4,36	1,86

in earlier studies [35]. The present research focus contributes to understanding the timeliness of functional programming [8] regarding humanitarian domain of interpretation. The obtained data paves way to further implementation of situational frame based-approach to tailor self-organized neural networks and to bottom-up generative processes [11] at both professional and educational levels. Moreover, one of the ways implies aggregation of data from English accent speech produced by various stakeholders [34], into particular speech banks tailored to specific communicative contexts. Finally, the above data confirms that Yandex Speech Interpretation system needs more improvement. The interim outcomes confirm the need for the research methodology development. Scholars underline the relevance of the deep latent factor model for high-dimensional and sparse matrices in recommender system [32]. Therefore, to ensure higher quality of the recommender platforms for SISs operation, unconstrained no-negative latent factor analysis (UNLFA) for speech-to speech recommender systems is critical due to its relevance in terms of prediction accuracy for missing data which can be introduced on grounds of those variables whose relevance was identified in the experiment, and compatibility of its unconstrained training process with various general training schemes [20]. Besides, to enhance prediction accuracy matrix factorization needs to use triple-factorization based approach [29] technique. The respective studies are supposed to be conducted at the next stage of the research.

3.2 Interpreters' Opinions

Professional interpreters were invited to consider the above videos data and discuss speech recognition and interpretation systems outputs. The discussion of the interpretation systems outputs with interpreters and their replies to the questionnaire made it possible to identify the clusters that characterize the scope of the discussion on AI interpretation systems relevance interpreters' point of view (Table 4).

Table 4. Data on speech interpretation systems use by interpreters, Author's data.

Clusters	% of those whose opinions formed the cluster
Reasons for speech interpretation systems use	81%
Contexts for speech interpretation systems use	81%
Target audiences for mediation via speech interpretation systems use	78%
Choice of specific speech interpretation systems for use	74%
Reasons for the above choice	75%
Satisfaction with the application of speech interpretation systems	71%

The replies also revealed some key features for the above cluster's identification.

The respondents mentioned a number of factors for reasons for SIS use (cluster 1). 1st factor referred to target audiences' poor capacity of' mediating language skills (English, French, German) (0,871), mentioned by 88% of the respondents who used with the above languages with no capacity to operate with the client's local languages. 2nd factor concerned interpreter's failure to understand the language of the interviewed due to its regional accent (0, 538), mentioned by 54% of the interpreters. 3rd factor revealed that SISs might be used for self-control tool during interpreting for people who used not a mother tongue (0,291). mentioned by 23% of the interpreters.

The second cluster identified major contexts where AI-based speech interpretation systems might be used. 1st factor identified mediation at the Russian borders crossing points (0,396), mentioned in 40% of replies). 2nd focused on interpreting for clients in outpatient clinics and hospitals (0, 458), mentioned in 50% of replies. 3rd factor referred to mediating at the police stations (0, 486), cited in 50% of replies. 4th factor covered interpretation at migration agencies (0, 401), mentioned in 40% of replies.

The third cluster specified target audiences for whom mediation might involve STS use. 1st factor referred to regular labor migrants (0, 368), cited by 37% of respondents. 2nd factor covered undocumented migrants (0, 628), cited by 63% of respondents.

The fourth cluster revealed interpreters' preferences regarding AI-based speech interpretation systems, including Google Translator (100% of interpreters confirmed), Yandex (17% of interpreters mentioned).

The fifth cluster elaborated on the above preferences and explained reasons for the above preferences. 1st factor explained reasons for system choice including the system open-source status and available language options regarding Google (0, 876), mentioned in 88% of the replies. 2nd factor concerned the need for specific realities search regarding Yandex in case of interpreting from and into Russian (0, 143), mentioned in 14% of the replies.

The six cluster concerned interpreters' satisfaction with the mentioned tools. Totally, 87% of the respondents were not satisfied with the service. 1st factor referred to cases when interpreters were not satisfied with the outputs due to failure in interpretation, errors in words and word combinations, (0,781), reported in 78% of the replies. 2nd factor covered those situations when there was no interpretation of the whole utterance (0, 231), reported in 23% of the replies.

Further on, the interpreters' free style comments on the items of the questionnaire (with respect to the comparative analysis of the three systems outputs within the present research procedure) were subject to conventional and summative content analysis. These types of study allowed the authors to code most frequent concepts that interpreters used to elaborate (Table 5).

Table 5. Content analysis of interpreters' views, Author's data

Content concept code	% of those who mentioned the concept
Situation components	81%
Identification of constituent concepts/actors	81%
Actors' verbal behavior	78%
Sentence structure	74%
Speech variation due to social variables	75%
Short simple language	71%
Questioned persons' potential errors forecast	69%

In their free-style comments, the respondents underlined that when tuning SISs, those engaged need to identify all possible scenarios for a particular situation, including topics, concepts, participants, their aspirations, etc. This statement confirms the importance of "boosting automatic event extraction" [34] for interpreting in unscheduled interpreting settings, emergencies, and conflicts. Earlier the need to explore the events constituent components was specified with respect to AI- supported communication in health care settings [30]. The mention of actors' verbal behavior implies that all possible forms of verbal expressions regarding a particular situation and its actor's sentence structures should be mapped. The focus on social variables underlines that individuals' variants and specifics of speech with respect to the gender, age, level of education, place of origin, professional and physical conditions require their inclusion in the speech bank for speech interpretation systems (the data may be taken from both existing banks and new real-world sources). The fact that interpreters mention these things enhances the list of meaningful variables, while earlier research mentioned emotional conditions [5]. Interpreters also pointed out that variations of utterances syntagmatic division, anaphora and ellipses might create problems for speech recognition systems regarding interpretation during police interviews, interviews with doctors. Such opinions add to the general understanding of the mentioned speech features [19] regarding the settings under study in the present research.

Moreover, some respondents connected their experience in speech interpretation systems use failures and deficiencies with diverse language prosodic patterns, as well.

The research findings make it possible to consider preliminary suggestions for training specialists who design SISs and use the technology under study for professional purposes, including border officers, emergency teams, social and healthcare workers, interpreters on site with no required working, language pair, etc. [18]. Specialists agree that the technology under study is used and will be used for communication with non-native speakers in emergency contexts due to lack of interpreters on-site [6]. Therefore, the technologies under study are supposed to be further developed. The Interpreters' voices should be heard, and users should be trained. The technology deficiencies are subject to IT specialists' activities which are supposed to improve the current SISs on grounds on the latest data from the field and users.

As far as the non-IT specialists are concerned, this category is expected to be aware of the existing SIS tools for multilingual communication, and possible challenges that the use of such tools might bring into cross language settings. This statement goes beyond the technical skills for the technology installation and operation. Here we imply the need for the development of specific communicative tactics, in terms of question and answers, as well as statements and instructions forms to be arranged in plain vocabulary and simplest grammar, self-control of the speech speed and articulation quality, behavior markers, etc. Specialists mention the importance of the listed features regarding humanitarian and emergency communication [12, 32] in general. To our mind, similar approach should be implemented to the use of SISs in multilingual communication in humanitarian/emergency settings.

4 Concluding Remarks

The study results confirmed that investigation of SISs capacity to serve communicative needs in emergency humanitarian contexts is well timed due to global migration and multilingual landscape increase. The research provided initial understanding of various SISs capacity, defined the types of errors and their average number per syntagma as the unit of sense in the communication. The empirical analysis contributed to understanding the needs for speech interpretation systems development regarding those contexts of cross-cultural communication that affect human rights and life and added new impetus to study of digitally supported interpretation from the angle of communicative failures, constrains and tensions management. The interpreters' surveys laid grounds to identify preliminary clusters characterizing possible angles for discussion on settings for SISs vital role in interlanguage communication. The interpreters' survey contributed to identifying those points that should be considered for interpreting systems learning and further tuning. The survey results put on the agenda the tasks to customize SISs to specific communicative settings in socially crucial contexts. Interpreters' survey also addressed the questions of those communicative features that should be taken in account while tuning digital interpretation systems to domains in terms of speech bank data to be included. Content analysis of interpreters' observations regarding the above systems tests shed light on points to be considered to design and improve the technology under study.

The respective issues seem to be subject for consideration within continuing professional development of specialists who work in the field. Tailored courses and bespoke training seem to be relevant for the agenda for training of those target audiences who deal with multilingual communication in emergency humanitarian settings. Moreover, a kind of module within the university-based course on speech-to-speech interpretation systems could be thought over for curriculum update.

The limitations of the present study stem from its pilot nature and research results interim status. Further studies require wider databases of language pairs, more specific communicative contexts, and further methodological advance.

APPENDIX. Links to Videos, Length of Recordings, Number of Syntagms (Last Accessed January 10, 2021)

Video 1. Refugee crisis: huge queues on the Serbian-Macedonian border (2015, November 11). https://www.youtube.com/watch?v=B43F3ZH-TTw (1.37–2.11; 2.44–3.25 min, 30 syntagms).

Video 2. Refugees face increasingly perilous journeys to Europe (2017, December 5). https://www.youtube.com/watch?v=Yma4OhnLLPw (1.46–2.12; 2.40–2.50 min, 10 syntagms).

Video 3. Rescued African migrants say they are fleeing slavery (2017, June 28). https://www.youtube.com/watch?v=lnSgWGUJ3jE (5.54–5.70; 6.23–6.44; 10.13–10.37 min, 24 syntagms).

Video 4. Surviving One of the Deadliest Routes to Europe: Refugees at Sea (2016, January 11). https://www.youtube.com/watch?v=nPelTu3iupc (3.36–5.12 min, 61 synt).

Video 5. Tensions between Afghan and Syrian refugees on the Greek island of Lesbos (2015, November 21). https://www.theguardian.com/world/video/2015/nov/21/tensions-between-afghan-and-syrian-refugees-on-the-greek-island-of-lesbos-video (0.41–0.52; 1.32–1.38; 1.50–3.03; 2.18–2.48; 3.53–4.13; 4.23–4.44; 5.32–5.48 min, 42 syntagms).

Video 6. Afghan refugees describe treacherous journeys to Turkey (2018, April 12). https://www.youtube.com/watch?v=yZACCMy2go8 (0.45–1; 1.07–1.20 min, 21 synt.)

Video 7. Niger refugees: Hundreds hope for a new life in Europe (2019, April 16). https://www.aljazeera.com/news/2019/04/niger-refugees-hundreds-hope-life-europe-190416104509532.html (0.17–0.43; 0.49–0.53 min, 55 syntagms).

Video 8. Unaccompanied refugee children share their dreams and despair … (2019, August 29). https://www.unicef.org/eca/stories-region/unaccompanied-refugee-children-share-their-dreams-and-despair-they-await-uncertain (0.46–1.17; 2.50–3.23; 4.02–4.59; 5.01–5.32; 5.44–7.31; 7.46–8.10, 9.40–9, 59–11.03–11.23 min, 106 syntagms).

Video 9. Rohingya English club conversation (2017, February 16) https://www.youtube.com/watch?v=G3rBY8N-3wE) (0–2.45 min, 38 syntagms).

Video 10. She tells her story on why she fled away from North Korea (2017, January 13 https://www.youtube.com/watch?v=EKbnyLKLHbo (6.37 min, 296 syntagms).

Video 11. From Myanmar to India (2019, December 3). https://thediplomat.com/2019/12/from-myanmar-to-india-refugees-lives-matter/ (6.19 min, 291 synt.).

References

1. Anumanchipalli, G.K., Chartier, J., Chang, E.F.: Speech synthesis from neural decoding of spoken sentences. Nature **568**, 493–498 (2019)
2. Bohouta, G., Këpuska, V.Z.: Comparing speech recognition systems (Microsoft API, Google API And CMU Sphinx). Int. J. Eng. Res. Appl. **7**(3), 20–24 (2017)
3. Chamchong, R., Wong, K.W. (eds): Multi-disciplinary Trends in Artificial Intelligence: 13th International Conference, Kuala Lumpur, Malaysia, 17–19 November 2019, Proceedings, vol. 11909. Springer, Cham (2019). https://doi.org/10.1007/978-3-030-33709-4
4. Chen, Zh.: Co-designing a chatbot for and with refugees and migrants. Unpublished master's thesis. Aalto University, Espoo, Finland (2019). https://aaltodoc.aalto.fi/handle/123456789/39282. Accessed 10 Jan 2021
5. Cominelli, L., Mazzei, D., De Rossi, D.E.: SEAI: social emotional artificial intelligence based on Damasio's theory of mind. Front. Robot. AI **5**, 6 (2018)
6. Česonis, R.: Human language technologies and digitalisation in a multilingual interpreting setting. In: Besznyák, R., Szabó, C., Fischer, M. (eds.) Fit-For-Market Translator and Interpreter Training in a Digital Age, pp. 179–195. Vernon Press, Wilmington (2020)
7. Dutta, S., Klakow, D.: Evaluating a neural multi-turn chatbot using BLEU score. Technical report. Saarland University, Saarbrücken (2019)
8. Flasiński, M.: Introduction to Artificial Intelligence. Springer, Cham (2016). https://doi.org/10.1007/978-3-319-58487-4
9. Fu, K.S.: Applications of Pattern Recognition. CRC Press, Boca Raton (2019)
10. Guest, G., Bunce, A., Johnson, L.: How many interviews are enough? An experiment with data saturation and variability. Field Methods **18**(1), 59–82 (2006)
11. Gonzalez-Rodriguez, D., Hernandez, R.: Self-Organized Linguistic Systems: from traditional AI to bottom-up generative processes. Futures **103**, 27–34 (2018)
12. Hunt, M., Pringle, J., Christen, M., Eckenwiler, L., Schwartz, L., Davé, A.: Ethics of emergent information and communication technology applications in humanitarian medical assistance. Int. Health **8**(4), 239–245 (2016)
13. Jackson, P.C.: Introduction to Artificial Intelligence. Courier Dover Publications, Mincola (2019)

14. Kandagal, A.P., Udayashankara, V.: Speaker independent speech recognition using maximum likelihood approach for isolated words. Int. J. Comput. Appl. **7**(6), 72–83 (2017)
15. Kim, J.B., Kweon, H.J., Lee, R. (eds.): Computational Science/Intelligence and Applied Informatics. CSII 2019. SCI, vol. 848, pp. 1–10. Springer, Cham (2020). https://doi.org/10.1007/978-3-319-96806-3
16. Kletečka-Pulker, M., Parrag, S., Droždek, B., Wenzel, T.: Language barriers and the role of Interpreters: a challenge in the work with migrants and refugees. In: Wenzel, T., Droždek, B. (eds.) An Uncertain Safety, pp. 345–361. Springer, Cham (2019). https://doi.org/10.1007/978-3-319-72914-5
17. Koenecke, A., et al.: Racial disparities in automated speech recognition. Proc. Natl. Acad. Sci. **117**(14), 7684–7689 (2020)
18. Lim, H.: Design for computer-mediated multilingual communication with AI support. In: Companion of the 2018 ACM Conference on Computer Supported Cooperative Work and Social Computing, pp. 93–96 (2018)
19. Luo, X., Zhou, M., Li, S., Wu, D., Liu, Z., Shang, M.: Algorithms of unconstrained non-negative latent factor analysis for recommender systems. IEEE Trans. Big Data **7**(1), 227–240 (2021)
20. Maučec, M.S., Brest, J.: Slavic languages in phrase-based statistical machine translation: a survey. Artif. Intell. Rev. **51**(1), 77–117 (2017). https://doi.org/10.1007/s10462-017-9558-2
21. Mishra, S.K.: Artificial Intelligence and Natural Language Processing. Cambridge Scholars Publishing, Cambridge (2018)
22. O'Brien, S., Federici, F., Cadwella, P., Marlowec, J., Gerberd, B.: Language translation during disaster: a comparative analysis of five national approaches. Int. J. Disast. Risk Reduct. **31**, 627–636 (2018)
23. Saad, U., Afzal, U.: El-Issawi: a model to measure QoE for virtual personal assistant. Multimedia Tools Appl. **76**(10), 12517–12537 (2016)
24. Shcherba, L.: Phonetics of the French Language, 7th edn. Higher School, Moscow (1963)
25. Shang, M., Luo, X., Liu, Z., Chen, J., Yuan, Y., Zhou, M.: Randomized latent factor model for high-dimensional and sparse matrices from industrial applications. IEEE/CAA J. Automatica Sinica **6**(1), 131–141 (2019)
26. Sinhababu, N., Saxena, R., Sarma, M., Samanta, D.: Medical information retrieval and interpretation: a question-answer based interaction Model. arXiv preprint arXiv:2101.09662 (2021)
27. Al Smadi, K., Al Issa, H.A., Trrad, I., Al Smadi, T.: Artificial intelligence for speech recognition based on neural networks. J. Signal Inf. Process. **6**(2), 66–72 (2015)
28. Strobel, M., Dwyer, C.: Obstacles to Adopting Speech Recognition in Emergency Services Solutions (2018). https://aisel.aisnet.org/amcis2018/AdoptionDiff/Presentations/25/
29. Song, Y., Li, M., Luo, X., Yang, G., Wang, C.: Improved symmetric and nonnegative matrix factorization models for undirected, sparse and large-scaled networks: a triple factorization-based approach. IEEE Trans. Industr. Inf. **16**(5), 3006–3017 (2020)
30. Wahl, B., Cossy-Gantner, A., Germann, S., Schwalbe, N.R.: Artificial intelligence (AI) and global health: how can AI contribute to health in resource-poor settings? BMJ Global Health **3**(4), e000798 (2018)
31. Wakatsuki, D., Kato, N., Shionome, T.: Development of web-based remote speech-to-text interpretation system captiOnline. J. Adv. Comput. Intell. Intell. Inform. **21**(2), 310–320 (2017)
32. Wu, Y., et al.: See What i'm saying? Comparing intelligent personal assistant use for native and non-native language speakers. In: 22nd International Conference on Human-Computer Interaction with Mobile Devices and Services, pp. 1–9 (2020)

33. Wu, D., Luo, X., Shang, Y., He, Y., Wang, G., Zhou, M.: A deep latent factor model for high-dimensional and sparse matrices in recommender systems. IEEE Trans. Syst. Man Cybernet. Syst. **51**(7), 4285–4296 (2021)

34. Xiang, W., Wang, B.: A survey of event extraction from text. IEEE Access **7**, 173111–173137 (2019)

35. Zhang, X., Miyaki, T., Rekimoto, J.: WithYou: automated adaptive speech tutoring with context-dependent speech recognition. In: Proceedings of the 2020 CHI Conference on Human Factors in Computing Systems, pp. 1–12 (2020)

Factors Affecting the Intention to Buy Online During Covid-19: Electronic Devices in Southern Vietnam

Nguyen Thi Phuong Giang[⊠], Vo Thi Huynh Han, Danh Thi Ngoc Anh, and Nguyen Binh Phuong Duy

Industrial University of Hcm City-Iuh, HCMC, Ho Chi Minh City, Vietnam
{nguyenthiphuonggiang,nguyenbinhphuongduy}@iuh.edu.vn,
{17030571.han,17029061.anh}@student.iuh.edu.vn

Abstract. E-commerce is an industry that has an important influence on businesses in building business strategies. Vietnamese businesses are now promoting the development of e-commerce transactions to create a new way of doing business and attracting more customers. The e-commerce market brings a lot of benefits to businesses as well as convenience to customers. However, the market of this industry has fierce competition not only between domestic competitors, but also foreign competitors. In particular, during the raging epidemic of acute respiratory infections of COVID-19, this competition is increasing. Knowing that, this study focuses on exploring the influential factors such as product value, perceived risk, website quality, trust and perceived usefulness that affect customers' online purchase intention. This study combines both qualitative and quantitative research methods. The survey has a scale of 481 people; the survey subjects are consumers in South Vietnam. After collecting the feedback samples, the data is analyzed using SPSS software. The results of the study show that product value, website quality, trust and perceived usefulness positively affect the online purchase intention. There is also a perceived risk factor that negatively affects online purchase intention. The study also showed that the three factors that have the greatest impact on online purchase intention are arranged in the following order: perceived usefulness, trust, and website quality. As a result of the study, a number of governance implications have been proposed to suit businesses in the context of the ongoing COVID-19 pandemic.

Keywords: Electronic products · Online purchase intention · Product value · Perceived usefulness · Perceived risk · Website quality

1 Introduction

The economy has been growing, people are increasingly busy, the time to shop as traditional is to go to stores, supermarkets are shrinking. Meanwhile, nowadays internet access means are increasingly popular, people can access by phone, laptop, desktop, other devices with internet connection and anywhere (at home, at work or on outings). It is for

M. A. Serhani and L.-J. Zhang (Eds.): SERVICES 2021, LNCS 12996, pp. 18–34, 2022.
https://doi.org/10.1007/978-3-030-96585-3_2

that reason that more and more people have formed the intention to buy online. Online shopping in the world is not new but is gradually accounting for a very high percentage of people's shopping, especially during the current COVID-19 pandemic. Currently, Vietnam is considered one of the fastest growing electronics markets in Southeast.

The E-Commerce market has been bringing many benefits to businesses and customers. However, the fierce competition of the online market is also a pressure on businesses. Therefore, to have a base, a development orientation and attract the interest of customers leads to increased consumption through websites. Businesses need to have an understanding and grasp what factors motivate consumers to buy. What factors affect the customer's purchasing intention? What prevents consumers from making purchases? This research is aimed at identifying factors affecting the intention to buy online during COVID-19: electronics in Southern Vietnam. From there, businesses build customer research models on the basis of premises and data analysis results. Help businesses and websites sell online in a timely manner to understand and adjust the policies of businesses so that more and more customers can make transactions.

2   Literature Review

The theory of planned behavior _ TPB was originally studied by Martin Fishbein and Ajzen in 1980. This is a model inherited from Theory of Reasoned Action _ TRA. TPB Theory Planned Behavior is successfully applied in many different fields. This theory refers to factors that affect the intended behavior and behavior of consumers, including: attitudes, subjective norms, perceived behavioural control. In particular, behavior control perception refers to human perception of how easy or difficult it is and the perception of controlling behavior can vary from situation to situation and action (Ajzen 1991, p. 183). Subjective attitude and norms are inherited from TRA theory. Attitudes to behavior and refers to the extent to which a person has a favorable or unfavorable assessment or assessment of the behavior in question (Ajzen 1991, p. 188). Subjective norms refer to the social pressure of being aware to perform or not to perform behavior (Ajzen 1991, p. 188). According to Ajzen (1991) argues that the more favorable the attitude and subjective norms for a behavior and the greater the ability to perceive behavioural control, the stronger the person should be the intention of an individual to perform the behavior under consideration.

The technology acceptance model _ TAM was introduced by Davis (1989). TAM refers more specifically to the prediction of the acceptability of an information system. The model aims to anticipate the application of certain information technology and propose necessary changes to information technology to achieve greater acceptance. Davis pointed to the influence of factors: Perceived usefulness and perceived ease of use. In the model, the perceived usefulness is defined as "the degree to which a person believes that using a particular system will enhance their work performance." The perceived ease of use is understood that the use of specific systems is easy and without difficulties (Davis 1989, p. 320). In Venkatesh and Davis (2000), the perception of usefulness is a strong decisive factor for intended use (p. 187).

2.1 Online Purchase Intention

The intention to make an online purchase is a trend of participating in online purchases or being willing to participate in consumer purchasing activities (Wen and Maddox 2013). In the study of Nguyen et al. (2019) Online shopping intention shows the extent to which customers intend to use e-commerce sites to shop in the future and can recommend others to do so online shopping practice. Online purchase intention is influenced by many factors. As in Sam and Tahir (2009) research, Purchase intention is influenced by factors such as: Usability, website design, Information quality, Trust, Perceived risk, Empathy. Or in the study Hemantkumar and his associates conducted in 2020 also showed that trust, perceived risk affects the intention to buy online, In addition, the study also showed other factors such as: perceived usefulness (PU), perceived ease of use, perceived behavioral control, E-shopping quality and subjective norms.

2.2 Website Quality

A quality website involves many different factors such as: Website design, site speed, content,... The design of websites plays an important role in attracting and retaining customers (Liao et al. 2006). Website design refers to the content on the site including: text, images, graphics, layout, sound, movement. These are identified as one of the main factors contributing to pulling customers back (Sam and Tahir 2009). Research on website design shows that providing richer media with a more realistic environment is more positive than the influence with user participation (Hausman and Siekp 2009). Some elements of web quality, such as information quality (Lederer et al. 2000; Lin and Lu 2000) have been verified as being related to Perceived Usefulness. Research on website design shows that providing richer media with a more realistic environment is more positive than the influence with user participation (Hausman and Siekpe 2009). The quality of the content of websites can increase consumer confidence in usefulness (Liao et al. 2006). From there the group of authors hypothesized, following:

H1: Website quality positively affects the intention to purchase electronic devices online.

H2: Website quality positively affects the perceived usefulness.

2.3 Trust

Trust is defined as the dimension of a business relationship that determines the extent to which each party feels they can rely on the integrity of the promise made by the other (Kolsaker and Payne 2002). Online shopping from traditional commerce and argue that trust is crucial for online trading (Chen et al. 2010). Trust is considered an important factor in affecting the purchasing intention of customers (Sam and Tahir 2009; Rasha Abu-Shamaa et al. 2015). In addition, trust is also believed to be weak which can increase the level of co-awareness usefulness (Gefen et al. 2003). From previous studies, the author gives two hypotheses as follows:

H3: Trust has a positive effect on the intention to purchase electronic online.

H4: Trust has a positive effect on the perceived usefulness.

2.4 Perceived Usefulness

Perception of usefulness is the powerful factor that determines the intended use of information technology (Venkatesh and Davis 2000; Zhu et al. 2012; Vijayasarathy et al. 2004; Hemantkumar et al. 2020). As well as previous research by Rasha Abu-Shamaa et al. (2015) colleagues on the issue, this study also suggested that perceived usefulness has a relationship with online shopping intention. It has a significant impact on online shopping intention. Many studies later gave similar results (García et al. 2020; Hemantkumar et al. 2020). Therefore, the author hypothesized as follows:

H5: Perceived usefulness positively affects the intention of purchasing electronic devices online.

2.5 Perceived Risk

Risk can be defined as the expectation of a defined loss of online consumers in a specific online purchase estimate (Hasan and Rahim 2008; Leeraphong and Mardjo 2013) There are many studies on the relationship between perceived risk and perceived usefulness. Such as Featherman and Wells (2010) found that perceived risk significantly reduced the perceived usefulness of paying bills online. Likewise, Li and Huang's findings (2009) show that the perceived risk has a negative effect on the perceived usefulness when shopping online. As Li and Huang (2009) claims that online shopping includes more uncertainty and risk than traditional shopping. According to research on student online purchasing intention has shown that risk-taking has a negative effect on online purchasing intention. In the Study Hemantkumar et al. (2020), also said that perceived risk is a factor in affecting consumers' online purchasing intention. Thus, the proposed hypothesis is:

H6: Perceived risk has a negative effect on the intention of purchasing electronic devices online.

H7: Perceived risk has a negative effect on perceived usefulness.

2.6 Product Value

Product value represents the perceived quality of products and services by consumers (Boyer and Hult 2006; Chen et al. 2010). It is possible that in online purchases the product value is focused on the quality of the product. However, reasonable price and high quality are equally important to increase the value of the product, thereby increasing the intention to buy (Turban et al. 2006). In this study the features that the author wanted to mention were based on the 2010 Study of Chen et al. including: features of the product that matched customer expectations (Boyer and Hult 2006; Brucks et al. 2000); easy-to-use products (Brucks et al. 2000); product price reflects reasonable product brand (Turban et al. 2006). The hypothesis given is:

H8: The value of the product has a positive effect compared to the intention of purchasing electronic devices online.

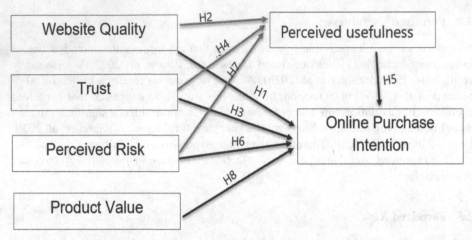

Fig. 1. Suggested author model (Source: Synthesis Author)

3   Research Method

The research was carried out with two methods: qualitative research and quantitative research (Zhu et al. 2012). Qualitative research is carried out by discussing the group through the phone. From the results of the group discussion, adjust the scale for the last time to make the official scale and the official questionnaire. Questionnaires for quantitative research are also established (Vijayasarathy 2004). The next step is quantitative research carried out with a total sample collection of 501 samples through online surveys and paper surveys. At the beginning of the cleaning phase, 20 imperfect samples were removed. Finally, the official data set used for the analysis was 481 and included in the statistics.

In terms of gender, the percentage of men who participated was 41.8% less than the female ratio. Meanwhile, the proportion of females accounted for 58.2%. In terms of age, with the age of 18–22 having 303 customers accounting for the highest proportion with 63.0%, the age of 23–27 years old participating in the survey accounted for nearly 16.0% 8%, the age of 28–32 years old participating in the questionnaire survey accounted for about 7.5%, the age of 33–37 years old participated in the survey accounted for nearly 8.9%, the age of over 37 years and older accounts for nearly 3.7%. Therefore, customers in Ho Chi Minh City have a high percentage of youth. In terms of careers, the percentage of students surveyed accounted for nearly 62.16% of the study samples. However, office workers also accounted for 16.01% of other occupations; civil servants and employees account for about 4.78% of other branches; with the profession of entrepreneur, managers participating in the survey accounted for nearly 2.29%; other professions surveyed by question accounted for nearly 14.76%. The survey was conducted in a variety of industries to help the study get a better overview of the intention to buy electronic devices. In terms of income, income of less than VND 3 million accounts for a high proportion of 40.1%. This was followed by income from 3 million to 7 million, accounting for 25.4%, income from 7 million–15 million accounting for 25.8% and finally the income of over 15 million accounted for the lowest 8.7% compared to other income levels.

Then the data were analyzed using SPSS software for EFA analysis and linear regression analysis. Finally, the research team discusses the research results and offers some management implications to improve the intention to buy electronic devices online for consumers in the South region. The detailed observed variables will be presented in Appendix 1.

4 Result

Cronbach's Alpha Test

Ensure sufficient reliability of the coals, conducting inspections with Cronbach's Alpha. The results showed that the reliability of all 7 scales had Cronbach's Alpha range from 0.788–0.895. In many previous studies, Cronbach's Alpha was between 0.8 and close to 1, the scale was good (quoted by Duy et al. 2020). From 0.7 to 0.8 is usable (quoted by Duy et al. 2020). Therefore, we conclude that the scales set out such as PV, PR, WQ, TRU, PU, OPI are standard and statistically significant. Details are presented in Table 1.

Table 1. Cronbach's alpha results for 6 coal measurements

No.	The scale	Number of observed variables	Coefficient cronbach's alpha
1	Product value (hereafter PV)	4	0.866
2	Perceived risk (hereafter PR)	4	0.788
3	Website quality (hereafter WQ)	3	0.895
4	Trust (hereafter TRU)	4	0.827
5	Perceived usefulness (hereafter PU)	3	0.892
6	Online purchase intention (hereafter OPI)	4	0.86

Explore Factor Analysis (EFA)

The study will perform EFA analysis, to ensure the value of the components in the suggestive element and discover other factors. This step will be performed EFA is conducted using the factor analysis performed using the "Principal Component" with the rotation "Varimax" as an extraction factor extracted Nesha et al. (2018). The results of the analysis are presented specifically in (Appendix 2).

After analyzing EFA for 5 independent variables, the result obtained a KMO value equal to 0.89 (conditions greater than 0.5 and less than 1) and Sig $= 0.00 < 0.05$ showed that the data was suitable for conducting factor analysis. The extracted variance is 72.417% ($>$50%) with this result showing that 5 extracted factors explain 72,417% of the fluctuations of the observed data. The Eigenvalues system is 1,008, the extraction system has good information significance. All observed variables have a load $>$0.5, so observation variables measure the concept we need to measure. The rotation component matrix shows the convergence of observation variables into factor groups. We see that all factors converge the right focus as stated in the summary of the scale.

EFA analysis for the dependent variable KMO value of 0.819 (conditions greater than 0.5 and less than 1), and Sig $= 0.00 <$0.05 show that the data is suitable for factor analysis. Extracted variance reached 70.384% ($>$50%) with this result showing that the 5 extracted factors explain 70.384% of the fluctuations of the observed data. The Eigenvalues system is 2,815 $>$ 1, theExtracted variance has good information significance.

So after analyzing the EFA discovery factor, the results showed that 5 independent factors were analyzed, in accordance with the theoretical basis and the model set out. Therefore the original research model will be retained as Fig. 1.

Regression Analysis

With the hypothetical model, the authors conduct regression analysis in turn according to the following steps:

Firstly, perform linear regression for the model to test the impact of independent variables (website quality, trust, perceived usefulness, perceived risk and product value). with the dependent variable (intention to purchase electronic devices online).

Second, performing a revoicing to verify the impact of three independent variables is the quality of the website quality, trust, and perceived risk with the dependent variable being the perceived usefulness.

1st Regression Analysis

To test the research hypothesis on the link between independent factors (PV, PR, WQ, TRU, PU) with dependent variables (OPI) - The intention to buy electronic devices online. Performing linear regression obtained the results and presented in the tables below:

Table 2. Model summary of factors that affect your intention to buy electronic devices online

Model	R	R^2	R_{adj}^2	Std. error of the estimate	Durbin-Watson
1	.816[a]	0.666	0.663	0.35814	1.992

[a]Predictors: (constant) PV, PR, WQ, TRU, PU.

Based on the table of results 2, we can give the following analysis:

The coefficient of determination R^2 indicates how much (%) of the dependent variable is explained by the independent variable. Table 2, the regression results show that the coefficient of determination $R^2 = 0.666$ ($\neq 0$) ie 66.6% of this index means that the variation of OPI variable is explained by the variation of 5 factors (PV, PR, WQ, TRU, PU), the remaining 33.4% belongs to other random factors and errors. In this model, using the $Radj^2$ index (R correction square) $= 0.663$ (66.3%) this index can help determine the model's fit more accurately and safely. The value of Durbin - Watson is 1,992 in the interval (1,3) from which it can be inferred that the data has no first-order correlation phenomenon, the data meets the requirements.

Table 3. ANOVA factors that affect the quality of relationships intended to purchase electronic devices online

	Sum of squares	df	Mean square	F	Sig.
Regression	121.705	5	24.341	189.775	.000[b]
Residual	60.925	475	0.128		
Total	182.63	480			

[b]Predictors: (Constant) PV, PR, WQ, TRU, PU.

Table 4. Regression weight table of factors affecting the intention to purchase electronic devices online

(Constant)	B	Std. error	β standard chemical	t	Sig.	Collinearity statistics		Hypotheses	Result
						Tolerance	VIF		
	1.047	0.116		9.005	0				
PV	0.151	0.024	0.205	6.365	0	0.679	1.472	H8	Accept
PR	−0.046	0.02	−0.06	−2.272	0.024	0.991	1.009	H6	Accept
WQ	0.151	0.026	0.208	5.775	0	0.539	1.855	H1	Accept
TRU	0.228	0.031	0.272	7.336	0	0.512	1.951	H3	Accept
PU	0.243	0.026	0.323	9.171	0	0.566	1.765	H5	Accept

Table 3 shows F = 189,775 and significant Sig. = 0,000 (sig. ≤ 0.05), which means that the recall model is consistent with the collected data and the included variables have statistical significance with 5%.

As presented in Table 4, the VIF ranges from 1,009–1855 < 2. Therefore, we can conclude that the multi-mutation phenomenon in this model is small. The Sig. of all PV, PR, WQ, TRU, PU variables is less than 0.05, so it is concluded that the above variables affect the intention to buy electronic devices online. Pv, WQ, TRU, PU variables have the same impact on variables depending on the intention to buy electronic devices online due to having a positive Beta system. Pr variables have an inverse impact on variables depending on the intention to buy electronic devices online due to having a positive Beta system.

From the analysis results, we have the regression model as:
$$OPI = 0.205PV - 0.06\,PR + 0.208WQ + 0.272TRU + 0.323OPI.$$

2nd Regression Analysis
The results of the second regression, to test the research hypothesis about the relationship between the independent factors, the independent variable (WQ, TRU, PR) and the dependent variable PU, the results are detailed in the Appendix. 3:

Regression results show that $F = 116,098$ and Sig. $= 0.000$ (sig. ≤ 0.05), which means that the regression model fits the collected data and the included variables have statistical significance with 5%. As the results presented in Appendix 3, the highest VIF index VIF $= 1,578 < 2$. Therefore, we can conclude that the phenomenon of multicollinearity in this model is small. Sig. of the variables WQ, TRU are all less than 0.05, so it can be concluded that the above variables have an influence on OPI. The variable PR has sig. $= 0.098 > 0.05$, so the PR variable does not affect the OPI variable. β standardized of the variables CLW $= 0.368$, STC $= 0.353$ that shows that both variables are positive numbers. So, these two variables have the same positive effect on the dependent variable OPI.

Test the Hypotheses of the Model
The proposed hypothesis consists of 8 hypotheses. Based on Table 4, the hypotheses (H1, H3, H5, H6, H8) all have Sig. < 0.05, so it can be concluded that the hypotheses H1, H3, H5, H6, H8 are accepted.. The remaining hypotheses H2, H4 and H7 are tested in the second regression, detailed in Appendix 3, showing that Sig. of the variables WQ(H2) and TRU(H4) are equal to 0.00 from which the conclusion is that the hypothesis H2 and H4 are accepted. The variable PR has Sig. $= 0.098 > 0.05$ from which the corresponding hypothesis H7 is rejected.

5 Discussion

Based on the results of the analysis above, the research team set out the topic of discussing factors affecting the intention to buy online during COVID-19: electronics in Southern Vietnam.

According to the linear regression equation after analysis, we can conclude that consumers' intention to buy electronic devices online in the South region, Vietnam is influenced by 5 main factors: product value, perceived risk, website quality, trust, and perceived usefulness. In which, the perceived risk factor has a negative impact on the intention to buy electronic devices online in the South region, Vietnam, the higher this factor, the lower the intention to buy electronic devices online. For example, when there are too many risks on a product such as quality risk, information security risk, etc. The higher the risk, the more it can disrupt the procurement process of electrical equipment customers online.

The factors of product value, website quality, trust and perceived usefulness have a positive impact on customers' intention to buy electronic devices online. The higher these factors, the higher the consumer's intention to buy electronics online. In particular, the perceived usefulness has the greatest impact, the second is the trust factor, the third is the website quality and finally the product value. As a result, the linear revoicing equation accurately reflects the correlation between independent variables and dependent variables in the research model.

The results achieved by the study are consistent with previous studies. The Hemantkumar et al. (2020) study found that perceived usefulness is the most influential factor in purchasing consumer electronics online. This study suggests that online retailers and online platform developers should focus more on the user experience.

It is also thought that trust is related to the element of online purchase intention. In the previous study (Leeraphong and Mardjo 2013; Heijden et al. 2005) gave similar results. These studies believe that trust has a positive effect on the purchasing intention of customers, so this factor needs to be considered and needs to improve consumer confidence in the service and products of the store itself.

Perceived risk is a factor that has a negative impact on the intention to buy electronic devices. This discovery is consistent with past studies (e.g., García et al. 2020; Hemantkumar and Pratiksinh 2020) In addition, "Another study results show that the risk of being aware is not related to perception of usefulness. This result is similar to those found in some past studies (e.g., Li and Huang 2009)" quotes García et al. (2020).

In addition, variables such as trust and web quality have an effect on perceived usefulness. The impact of web quality significantly affects users who are aware of usefulness in accordance with studies by Heijden et al. (2005) and Liao et al. (2006). Trust can increase the level of perceived usefulness, the results of which are implied by the study of Gefen et al. (2003) and Liao et al. (2006).

6 Conclusion

From the results of the analysis and discussion of the study, the authors team came to conclusions. The impact of factors on the purchasing intention of customers is arranged in the following order: perceived usefulness, trust, website quality, perceived risk and product value. In particular, the perceived risk factor has a negative impact on the online purchase intention of customers. On the basis of the results of the study "propose solutions suitable for online business for businesses". From there, set out business methods and marketing strategies as well as other activities to meet the needs of customers. This helps to boost business efficiency, increase revenue for businesses and dominate the market.

In order to make it more convenient for businesses to develop business strategies for company development, the research team has given administrative implications on the factors of perceived usefulness, trust and website quality.

Perceived usefulness is the factor that most influences consumers' intention to purchase electronic products online. This implies that online entrepreneurs should focus on the customer experience more. Because the survey results of this study show that customers want to experience good customer care and can communicate with sellers as quickly as possible, they want to save more time in shopping. Managers need to come up with business strategies, allocate rich products and focus on providing true, accurate, detailed and timely information about products that meet the needs of customers. In addition, the quality of the site also affects the perception of usefulness. Therefore, the quality of service of the online platform is also a factor that needs attention.

The unpredictable nature of internet infrastructure is growing, consumers are concerned that hackers or third parties will threaten their financial secrets or disclose personal information (Featherman and Pavlou 2003 quoted in Liao et al. 2006). Therefore, electronics providers need to enhance consumers' online purchase intentions by enhancing their trust. Specify and provide rules and regulations for online trading. From there, it is possible to make customers more confident and reliable in online goods. In addition, increased customer care before and after purchase, is essential in the e-commerce environment. According to the survey, responding to online questions is a problem that customers evaluate as the factor that has not yet achieved the highest scale in this e-commerce environment.

Through the results of the analysis, the research team pointed out the quality factor of the website quality influential in the intention to buy electronic devices online in Ho Chi Minh City. This study also shows that in order to develop a consumer-oriented e-commerce business, administrators need to focus on the following: Focusing on the quality of information about electronic products on online stores. The content of the information on the website must be accurate, complete, clear and reliable. Online stores should focus on the necessary information for customers to refer and choose products accordingly from Tran Anh Vu et al. (2020). In particular, administrators need to improve the quality and speed up website access so that customers can shop easily, avoiding the case of not being able to connect to reduce the intention to buy or interrupt the shopping needs of customers. The adjustment of the business strategy of the business/company according to the above results to meet the needs of each customer.

Although in the course of implementation there have been many attempts to complete the study, like with some other studies, this study also has some limitations. Due to limited time spent doing the research, the team did not analyze the variables in depth. The research by the authors has only recently tested consumers' intention to buy electronic devices online in the South region, Vietnam, so the level of generalization may not be high. Therefore, the team proposes to expand the scope of research in many different provinces and the next study should compare which key factors affect the online purchase intention of electronic devices between countries. Finally, the authors suggest that future studies should further explore other factors or attributes that may influence the intention to purchase electronic devices online.

Appendix

Appendix 1 Summary of observed variables

Table 5. Summary of observed variables

Factor	Scale	Code	Source
Product value	The features of the product meet your requirements	PV1	Chen and ctg (2010)
	The online store sells good quality products	PV2	
	Online store selling products at reasonable prices	PV3	
	The online store sells products of clear origin and guaranteed genuine	PV4	
Perceived risk	You find it difficult to accurately assess product quality when shopping online	PR1	Hemantkumar et al. (2020)
	You find it very difficult to compare the quality of similar products when shopping online	PR2	
	Personal information such as your address, email, phone number may be disclosed to others	PR3	
	You find your shopping habits and process easy to track when shopping online	PR4	
Website quality	Product information and images are detailed and clear	WQ1	Hemantkumar et al. (2020), Chen et al. (2010)
	The online store has a user-friendly interface designed to be easy to use	WQ2	

(*continued*)

Table 5. (*continued*)

Factor	Scale	Code	Source
	The website has a fast loading speed to help you find the exact product in a short time	WQ3	García et al. (2020)
Trust	The online store has a solid confirmation message when you close the purchase	TRU1	Chen et al. (2010), Hwang and Kim (2007)
	The website's online reply service meets your requirements	TRU2	Hemantkumar et al. (2020), Chen et al. (2010), Hwang and Kim (2007)
	The terms of the transaction (including payment, shipping, warranty, return, etc.) are detailed and clear by the online store	TRU3	
	The online store is known by many to be reputable and trustworthy	TRU4	
Perceived usefulness	An online store that offers a wide range of electronic products information	PU1	Hemantkumar et al. (2020), Gefen et al. 2003), García et al. (2020)
	Online store and you can easily exchange information back and forth	PU2	
	Buying online saves customers shopping time	PU3	
Online purchase intention	It is likely that you will purchase electronic products through an online store	OPI1	García et al. (2020), Hausnam and Siepe (2009)
	You will definitely buy electronic products through online stores	OPI2	
	You will introduce other customers to buy electronic devices online	OPI3	
	You will continue to purchase electronic products through the online store	OPI4	
Hypothesis: 6; Number of variables observed: 22			

Appendix 2

Table 6. EFA for 5 independent variables

Rotated component matrix

	Component				
	1	2	3	4	5
PV2	0.868				
PV1	0.825				
PV3	0.771				
PV4	0.706				
TRU2		0.778			
TRU3		0.735			
TRU4		0.712			
TRU1		0.569			
PR3			0.791		
PR4			0.76		
PR2			0.758		
PR1			0.712		
PU3				0.859	
PU2				0.823	
PU1				0.74	
WQ2					0.847
WQ1					0.791
WQ3					0.786
Cronbach alpha times 2	**0.866**	**0.827**	**0.788**	**0.892**	**0.895**

Table 7. The result obtained a KMO value

KMO	0.89
Bartlett's (sig.)	0.00
Method wrong deduction	72.417%
Eigenvalue	1.008

Table 8. EFA for OPI dependent variable

Component matrix	
	Component
	1
OPI2	0.863
OPI4	0.855
OPI3	0.819
OPI1	0.818
KMO	0.819
Bartlett's (sig.)	0.00
Method wrong deduction	**70.384%**
Eigenvalue	**2.815**

Appendix 3

Table 9. Regression analysis

Coefficients[a]									
Model						Collinearity statistics			
		B	Std. error	beta	T	Sig.	Tolerance	VIF	Result
1	(Constant)	0.956	0.194		4.94	0			
	WQ	0.356	0.042	0.368	8.426	0	0.635	1.575	Accept
	TRU	0.395	0.049	0.353	8.074	0	0.634	1.578	Accept
	PR	0.058	0.035	0.058	1.659	0.098	0.996	1.004	Rrefute

[a]Dependent variable: OPI.

References

Hausman, A.V., Siekpe, J.E.: The effect of web interface features on consumer online purchase intentions. J. Bus. Res. **62**, 5–13 (2009)

Gefen, D., Karahanna, E., Straub, D.W.: Trust and TAM in online shopping: an integrated model. MIS Q. **27**(1), 51–90 (2003)

Zhu, D.-S., Lin, T.-T., Hsu, Y.-C.: Using the technology acceptance model to evaluate user attitude and intention of use for online games. Total Qual. Manag. Bus. Excell. **23**(7–8), 965–980 (2012)

Vijayasarathy, L.R.: Predicting customer intention to use on-line shopping: the case for an augmented technology acceptance model. Inf. Manag. **41**(6), 747–762 (2004)

Venkatesh, V., Davis, F.D.: A theoretical extension of the technology acceptance model: four longitudinal field studies. Manage. Sci. **46**(2), 186–204 (2000)

Hasan, H.H., Rahim, S.A.: Factors affecting online purchasing behavior. Malaysian J. Commun. **24**, 1–19 (2008)

Wen, G.R., Maddox, S.L.: Influence factors on users' online shopping in China. Asia Bus. Mag. **7**(3), 214–230 (2013)

Nguyen, T.H., Long, T.T.V., Thuy, P.T., Anh, L.T.T.: Online shopping intention and behavior of customers: study of the extended unified theory of acceptance and use of technology, trust and long tail effect. VNU J. Sci. Econ. Bus. **35**(1) 112–120 (2019)

Ajzen, I.: The theory of planned behavior. Organ. Behav. Hum. Decis. Process. **50**, 179–211 (1991)

Bulsara, H.P., Vaghela, P.S.: Online shopping intention for consumer electronics products: a literature review and conceptual model. E-Comm. Fut Trends **7**(1), 24–32 (2020)

Sam, M., Tahir, H.: Website quality and consumer online purchase intention of air ticket. Int. J. Basic Appl. Sci. IJBAS **9**(2009)

Liao, C., Palvia, P., Lin, H.-N.: The roles of habit and web site quality in E-commerce. Int. J. Inf. Manag. **26**(6), 469–483 (2006)

Lederer, A.L., Maupin, D.J., Sena, M.P., Zhuang, Y.: The technology acceptance model and the World Wide Web. Decis. Supp. Syst. **29**(3), 269–282 (2000)

Lin, J.C., Lu, H.: Towards an understanding of the behavioral intention to use a web site. Int. J. Inf. Manag. **20**(3), 197208 (2000)

Abu-Shamaa và cộng sự, R.: Factors influencing the intention to buy from online stores: an empirical study in Jordan. In: Proceedings of the 8th IEEE GCC Conference and Exhibition, Muscat, Oman, 1–4 February, 2015

Leeraphong, L., Mardjo, A.: Trust and risk in purchase intention through online social network: a focus group study of facebook in thailand. J. Econ. Bus. Manag. **1**(4) (2013)

Featherman, M.S., Wells, J.D.: The intangibility of e-services: effects on perceived risk and acceptance. ACM SIGMIS Database **41**(2), 110–131 (2010)

Li, Y., Huang, J.: Applying theory of perceived risk and technology acceptance model in the online shopping channel. World Acad. Sci. Eng. Technol. **53**, 919–925 (2009)

Turban, E., King, D., Lee, J.K., Viehland, D.: Electronic Commerce 2006: a Managerial Perspective. Prentice Hall, Upper Saddle River (2006)

Duy, B. P. N., Bình, L.C., Giang, N.T.P.: Factors affecting students' motivation for learning at the industrial University of Ho Chi Minh City. In: Proceedings of the Future Technologies Conference (FTC) 2020, vol. 2, pp. 239–262 (2020)

Nesha, A.U., Rashed, M.S., Raihan, T.: Identifying the factors that influence online shopping intentions and practices: a case study on Chittagong Metropolitan City. The Comilla Univ. J. Bus. Stud. Bangladesh **5**(1), 157–171 (2018)

Van der Heijden, H., Verhagen, T., Creemers, M.: Understanding online purchase intentions: contributions from technology and trust perspectives. Eur. J. Inf. Syst. **12**(1), 41–48 (2005)

Featherman, M.S., Pavlou P.A.: Predicting e-services adoption: a perceived risk facets perspective. Int. J. Hum.-Comput. Stud. **59**(4), 451–74 (2003)

Anh Vũ, T., et al.: Customer's behavior on intention to purchase on online shopping in Vietnam. J. Bus. Manag. Sci. **8**(3), 85–88 (2020)

García, P., et al.: Purchase intention and purchase behavior online: a cross-cultural approach . Heliyon **6**(6), e04284 (2020)

Chen, Y.-H., et al. Website attributes that increase consumer purchase intention: a conjoint analysis. J. Bus. Res. **63** (2010)

Davis, F.D.: perceived usefulness, perceived ease of use, and user acceptance of information technology. MIS Q. **13**(3), 319–340 (1989)

Brucks, M., Zeithaml, V.A., Naylor, G.: Price and brand name as indicator of quality dimensions of consumer durables. J Acad. Mark. Sci. **25**(2), 139–153 (2000)

Boyer, K.K., Hult, G.T.M.: Customer behavioral intentions for online purchases: an examination of fulfillment method and customer experience level. J. Oper. Manag. **24**(2), 124–147 (2006)

Kolsaker, A., Payne, C.: Engendering trust in e-commerce: a study of Gunder-based concerns. Mark. Intell. Plan. **20**(4), 206–214 (2002)

Research on the Impact of Producer Services Industry Agglomeration on the High Quality Development of Urban Agglomerations in the Yangtze River Economic Belt

Jinhong Xu[✉] and Yi Li

ShenZhen Institute of Information Technology, Shenzhen 517000, China

Abstract. This paper uses the green total factor productivity (GTFP) to measure the level of high-quality economic development, and uses the partial Durbin model and Partial differential method for spatial regression model to explore the impact of producer service industry agglomeration on high-quality economic development and its sources. This paper analyzes the heterogeneity effect of producer services specialized agglomeration and diversified agglomeration on high-quality economic development in Chengdu Chongqing three urban agglomerations, and empirically tests the local direct effect and intercity indirect effect of producer services agglomeration on high-quality economic development in urban agglomerations. Based on the empirical results, this paper discusses the mode selection and policy recommendations of industrial agglomeration of producer services to promote the high-quality economic development of urban agglomeration.

Keywords: Producer services industry agglomeration · Urban agglomeration · High quality economic development · Green Total Factor Productivity (GTFP)

1 Introduction

The research on the relationship of economic development between geographically adjacent cities can be traced back to the theories of "echo diffusion", "polarization trickle", "core edge" of scholars such as Muldahl, Hirschman and Friedman. According to these theories, the influence of big cities on the surrounding small cities will gradually change from "siphon effect" to "diffusion effect" with the economic development stage; However, the theory of "agglomeration shadow" proposed by new economic geography holds that the surrounding cities of core cities are mainly affected by the negative effect of "siphon effect".

The research on the Yangtze River Delta Urban Agglomeration by Sun et al. (2016) and the research on the Pearl River Delta Urban Agglomeration by Mei et al. (2012)

Fund Project: ShenZhen Institute of information technology, school level doctoral program of Social Sciences–Research on the high quality development of ShenZhen industrial system under the new development pattern of double cycle, SZIIT2021SK010.

M. A. Serhani and L.-J. Zhang (Eds.): SERVICES 2021, LNCS 12996, pp. 35–52, 2022.
https://doi.org/10.1007/978-3-030-96585-3_3

show that big cities can drive the economic growth of surrounding cities, that is, there is "spillover effect"; Sun et al. (2013), Wang et al. (2014) and Zhang et al. (2016) found that the impact of core cities on the economic growth of surrounding cities is "polarization effect"; Chen and Padridge (2013) analyzed the interaction between Chinese cities from the perspective of urban hierarchy, and concluded that mega cities did not have a significant impact on county-level cities.

The Yangtze River economic belt is a typical representative of China's urban agglomeration economic belt, which spans 11 provinces and cities, including the Yangtze River Delta, the middle reaches of the Yangtze River, and Chengdu and Chongqing urban agglomeration. The population and GDP account for more than 40% of the country. However, the unbalanced development of the urban agglomeration in the Yangtze River economic belt is obvious, the situation of resources and environment is grim, and the input-output differences are large.

Producer service industry is a service industry that provides intermediate input for the production of other products or services. It has the characteristics of knowledge intensive, strong specialization, industrial integration and high innovation activity. It is the strategic commanding point of a new round of global industrial competition. By affecting the green total factor productivity (GTFP)[1], to promote the high-quality development of regional economy. By sorting out the current research literature, it is found that there are few research results on how to improve the regional green total factor productivity from the perspective of producer services agglomeration, and most literatures ignore the spatial spillover effect of explanatory variables on GTFP.

The economic essence of the development of urban agglomeration is the effect of agglomeration economy. Producer service industry is an important starting point for the transformation and upgrading of China's current industrial structure. It is of great practical significance to study the impact of its different agglomeration modes on the high-quality economic development of the three major urban agglomerations in the Yangtze River economic belt, as to promote the green and efficient development of the Yangtze River economic belt. In view of the uncertainty of spatial spillover effect of agglomeration economy in urban agglomerations, this paper will compare the green total factor productivity growth effect of producer services agglomeration in three urban agglomerations of the Yangtze River economic belt, so as to provide a basis for exploring a new path of high-quality economic development of urban agglomerations.

The innovation and marginal contribution of this paper are as follows: Firstly, in terms of research content, GTFP which considering resource and environment constraints is adopted to measure the level of high-quality economic development, and the impact and source of producer services agglomeration on high-quality economic development in urban agglomeration are directly discussed; Secondly, from the perspective of research, this paper distinguishes the local effect and intercity effect of producer services agglomeration, compares and analyzes the heterogeneous effects of specialized agglomeration and diversified agglomeration of producer services on high-quality economic development in the Yangtze River Delta, the middle reaches of the Yangtze

---

[1] GTFP considers "energy consumption" and "environmental pollution", Scientific integration of innovation driven and green development concept, relative to total factor productivity (TFP), can better reflect the requirements of high-quality economic development in the new era.

River and Chengdu and Chongqing urban agglomerations, and then explores the mode selection of industrial agglomeration under the goal of high-quality economic development of urban agglomerations; Thirdly, in order to identify the spatial spillover effect of producer services industry agglomeration among cities, this paper empirically tests the local direct effect and intercity indirect effect of producer services agglomeration on the high-quality economic development of urban agglomeration by using the partial Durbin model and Partial differential method for spatial regression model, It overcomes the shortcomings of traditional measurement method which ignores the interaction between cities.

2 Model Construction

2.1 The Setting of Space Weight Matrix

Since the transaction costs such as commuting cost and transportation cost caused by geographical distance play an important role in the economic connection between cities, this paper selects the weight of geographical distance as the weight matrix wij of spatial econometric model. The production transaction of urban producer services industry involves invisible knowledge and intensive and complex contractual arrangements. The spread of invisible knowledge decreases sharply with the increase of geographical distance. A shorter geographical distance is conducive to the establishment of trust relationship, the reduction of information asymmetry, and the increase of access to non-standard information. As shown in Eq. (1), this paper uses the exponential distance attenuation function to construct the spatial weight matrix wij, where d is the spherical distance between cities.

$$w_{ij} = \begin{cases} e^{-\alpha d_{ij}}, & i \neq j \\ 0, & i = j \end{cases} \tag{1}$$

2.2 Moran's I Index

The global spatial correlation of the green TFP index in the three major urban groups, namely Yangtze River Delta, middle Yangtze River and Chengdu and Chongqing, is tested separately by using global Moran's I index, as shown in formula (2). The value range of Moran's I index is between -1 and 1. The absolute value can indicate the global spatial correlation degree. The larger the absolute value, the stronger the spatial correlation degree of the high-quality development level of economy between cities. Where, $\bar{y}$ and S^2 represent the mean and variance of GTFP index, n is the number of cities in each urban agglomeration, w_{ij} is the spatial weight matrix set already, and y_i is the GTFP index of the ith city.

$$\text{Moran's I} = \frac{n \sum_{i=1}^{n} \sum_{j=1}^{n} w_{ij}(y_i - y)(y_j - y)}{(\sum_{i=1}^{n} \sum_{j=1}^{n} w_{ij} \sum_{i=1}^{n} (y_i - y)^2} = \frac{\sum_{i=1}^{n} \sum_{j=1}^{n} w_{ij}(y_i - \bar{y})(y_j - \bar{y}))}{S^2 \sum_{i=1}^{n} \sum_{j=1}^{n} w_{ij}} \tag{2}$$

2.3 Setting of Spatial Panel Econometric Model

In order to ensure the robustness of the econometric test results, according to the different sources of spatial correlation, this paper first uses three spatial econometric models: spatial lag model (SAR), spatial error model (SEM) and spatial Durbin model (SDM) for regression analysis under the spatial geographic weight matrix, and then compares the log likelihood and R^2 values, Choose to use the spatial Durbin model to build the SDM model as shown in Eq. (3).

$$\ln GTFP_{it} = \alpha + \rho \sum_{j=1}^{n} w_{ij} GTFP_{jt} + \beta X_{it} + \theta \sum_{j=1}^{n} w_{ij} x_{it} + \mu_i + \lambda_t + \varepsilon_{it}$$
$$\varepsilon_{it} = \emptyset \sum_{j=1}^{n} w_{ij} \varepsilon_{jt} + \varepsilon_{it} \tag{3}$$

Where i and t represent the city and year respectively, j represents other cities in the same urban agglomeration, and $GTFP_{it}$ represents the high-quality economic development level, β is the coefficient matrix, x_{it} is the explanatory variable vector, w_{ij} is the spatial weight matrix, μ_i and λ_t is the non observed effect of area and time, ε_{it} is the random disturbance term. $\rho \sum_{j=1}^{n} w_{ij} GTFP_{jt}$ is a spatial lag term, which indicates the influence of the GTFP of j city on the GTFP of i city, ρ and $\emptyset$ represents the spatial dependence of the explained variable and the error term respectively. Similarly, θ is to explain the coefficient of spatial lag term of variable x_{it}.

2.4 Decomposition of Spatial Spillover Effect of Independent Variables

According to LeSage and Pace (2009), the spatial spillover effect of independent variables is decomposed into direct effect and indirect effect.

$$y = \alpha l_n + \rho W_y + \beta X + \theta W X + \varepsilon \tag{4}$$

$$(l_n - \rho W) y = \alpha l_n + \rho W_y + \beta X + \theta W X + \varepsilon \tag{5}$$

$$y = \sum_{r=1}^{k} S_r(W) x_r + V(W)_{l_n \alpha} + V(W) \varepsilon \tag{6}$$

$$\begin{Bmatrix} y_1 \\ y_2 \\ \vdots \\ y_n \end{Bmatrix} = \sum_{r=1}^{k} \left\{ \begin{bmatrix} S_r(w)_{11} & S_r(w)_{12} & \cdots & S_r(w)_{1n} \\ S_r(w)_{21} & S_r(w)_{22} & & S_r(w)_{2n} \\ \vdots & & \ddots & \vdots \\ S_r(w)_{n1} & S_r(w)_{n2} & \cdots & S_r(w)_{nn} \end{bmatrix} \right\} \begin{Bmatrix} x_{1r} \\ x_{2r} \\ \vdots \\ x_{nr} \end{Bmatrix} + V(W)_{l_n \alpha} + V(W)_i \varepsilon \tag{7}$$

$$y_i = \sum_{r=1}^{k} [S_r(W_{i1} x_{1r} + S_r(W_{i2} x_{2r} \ldots + S_r(W_{in} x_{nr})] + V(W)_{l_n \alpha} + V(W)_i \varepsilon \tag{8}$$

$$\frac{\partial y_i}{\partial x_{jr}} = S_r(W)_{ij} \tag{9}$$

$$\frac{\partial y_i}{\partial x_{ir}} = S_r(W)_{ii} \tag{10}$$

Where $S_r(W) = V(W)(I_n\beta_r + W\theta_r)$, $V(W) = (I_n - \rho W)^{-1} = I_n + \rho W + \rho^2 W^2 +$ ..., I_n is the identity matrix of order n, ι_n is the identity matrix of order N * 1; x_r is the explanatory variable, $r = 1, 2,..., k$, β_r is the regression coefficient of the explanatory variable, θ_r is the regression coefficient of WX. $S_r(W)_{ii}$ is the direct effect, which is the effect of x_r of city i on the dependent variable of the city; $S_r(W)_{ij}$ is the indirect effect, which is the effect of x_r of city j on the explained variable of city i. The total effect is the sum of local direct effect and intercity indirect effect. The direct effect represents the influence of explanatory variables on the high-quality development of local economy, while the indirect effect represents the influence of independent variables of neighboring cities on the high-quality development of local economy in urban agglomeration.

3 Variables Setting

At present, most literatures use provincial level data to study from the perspectives of foreign direct investment, environmental regulation, energy structure, industrial structure and R & D investment. Based on the purpose of research, this paper focuses on the spatial spillover effect of producer services industry specialization and diversification agglomeration on high-quality economic development. As the above literature review has shown that regional GTFP is also affected by environmental rules, urban infrastructure, human capital, foreign investment participation and government intervention, these variables need to be introduced into the spatial econometric model as control variables. The specific variables setting and calculation method are shown in Table 1 below.

3.1 High Quality Economic Development Indicators

This paper uses GTFP to measure the high-quality economic development. Combining the unexpected output SBM model and super efficiency SBM model, this paper constructs the unexpected output super efficiency SBM model, and uses the GML index to measure the urban GTFP. Labor input, capital input and energy input are respectively measured by the total number of urban employees, capital stock of fixed assets and the city's annual electricity consumption. The capital stock is estimated by perpetual inventory method, and the calculation formula is Kit = $(1 - \delta)$ Ki, t $- 1 + $ Iit, i and t denote regions and years, K and I denote capital stock and new social fixed asset investment respectively, δ represents the depreciation rate of fixed assets. The total investment in fixed assets of the whole society of the city is selected as the investment index, and the price index of fixed assets investment of the province where the city is located is used for adjustment. Using the algorithm of Shan (2008), the depreciation rate is set at 10.96%. For the calculation of the capital stock in the base period, we refer to Hall and Jones (1999), use the sum of the average annual growth rate and depreciation rate of the actual gross fixed capital formation to calculate. The formula is: K0 = I0/(gi) $+ \delta$), gi is the aggregate average growth rate of real investment over a period of time.

The expected output index is expressed by the city's real GDP. Taking 2003 as the base period.

According to Tu (2008) and other relevant research on provincial level, the unexpected output index takes the city's industrial SO2 emissions, industrial waste-water emissions and industrial dust emissions.

Table 1. Variables setting and calculation method

Variables		Calculation method
Explained variable	High quality economic development indicators (GTFP)	Urban GTFP index based on SBM super efficiency model and GML index method
Core explanatory variables	Specialized agglomeration of producer services industry (SP)	Relative specialization agglomeration index of producer services industry
	Diversified agglomeration of producer services (DV)	Relative diversification agglomeration index of producer services industry
control variable	Information infrastructure level (Inf)	Internet access per 10000 people
	Transport infrastructure level (Road)	Urban Road area per person (m^2)
	Human capital level (Hum)	Number of students in Institutions of higher learning per 10000 people
	Foreign investment participation (Fdi)	The proportion of foreign capital actually used in cities in GDP
	The degree of government intervention (Gov)	Proportion of fiscal expenditure in fiscal revenue
	Environmental regulation (WJ)	Removal rate of SO$_2$ in urban industry

3.2 Core Explanatory Variables

The core explanatory variable of this paper is producer services industry agglomeration. In order to overcome the influence of scale factor, Puga (1999) method is used to measure the agglomeration level of producer services by using relative specialization agglomeration index (SP) and relative diversification agglomeration index (DV).

$$Specialization\ agglomeration\ index: SPi = maxjsij/Sj \qquad (11)$$

$$Diversity\ clustering\ index: DVi = 1/\sum j|Sij - Sj| \qquad (12)$$

Among them, Sij is the employment share of producer service industry j in the total number of urban employees, and Sj is the employment share of producer service industry j in the total number of urban employees in China. Referring to the research of Xi et al. (2015), the service industry with intermediate demand rate greater than 60% is defined as productive service industry.

According to the employment statistics of different industries in Chinese cities, six industries are defined as productive service industry, namely "transportation, warehousing and postal industry", "information transmission, computer service and software industry", "wholesale and retail industry", "financial industry", "leasing and business service industry" and "scientific research and technical service industry".

3.3 Control Variables

Environmental regulation (WJ): The industrial sulfur dioxide removal rate of each city is used to measure the level of environmental regulation.

Transport infrastructure level (Road): The per capital Road area is used to measure the level of transportation infrastructure.

Information infrastructure level (Inf): the number of Internet access per 10000 people to reflect the level of urban information infrastructure.

Human capital level (Hum): The number of students in secondary and higher schools per 10000 people to measure the level of urban human capital.

Foreign investment participation (Fdi): The proportion of the annual actual amount of foreign investment in GDP, and converts the amount of foreign investment according to the average price of RMB exchange rate over the years.

Degree of government intervention (Gov): The proportion of fiscal expenditure in fiscal revenue to express the degree of local government intervention in urban economy.

4 Empirical Test and Result Analysis

4.1 Data Sources

The research sample selected in this paper is the panel data of 71 cities in the three major urban agglomerations of the Yangtze River economic belt. Among them, the Yangtze River Delta Urban Agglomerations include 26 cities, namely Shanghai, Nanjing, Wuxi, Changzhou, Suzhou, Nantong, Yancheng, Hangzhou, etc.; There are 29 cities in the middle reaches of the Yangtze River, including Wuhan, Huangshi, Ezhou, Huanggang, Xiaogan, Xianning, Xiantao, Qianjiang; Chengdu and Chongqing urban agglomeration includes 16 cities, namely Chongqing, Chengdu, Zigong, Luzhou, Deyang, Mianyang etc. The original data used in this paper are mainly from the "China City Statistical Yearbook" and "China Statistical Yearbook" from 2003 to 2017. Individual missing data are filled with interpolation method, and the relevant variables are logarithmically processed.

4.2 Basic Assumptions

Hypothesis 1: the GTFP index of cities in urban agglomerations has a positive spatial correlation; Hypothesis 2: the high-quality development level of urban economy in urban agglomeration is affected by the agglomeration of local producer services industry, and also by the spatial spillover of producer services industry agglomeration in other cities in urban agglomeration; Hypothesis 3: the specialization and diversification agglomeration of producer services industry in urban agglomeration affect high-quality development, and the indirect effect of agglomeration on high-quality development is greater than 0.

4.3 Empirical Results

MaxDEA7 Ultra software is used to measure the GTFP index of 71 cities in the three major urban agglomerations of the Yangtze River economic belt.

4.3.1 Accounting Results of GTFP Index of Three Urban Agglomerations

Table 2 summarizes the annual mean of GTFP index of the three urban agglomerations. The results show that: first, from the geometric mean of the GTFP index of urban agglomeration over the years, the Yangtze River Delta urban agglomeration has the largest improvement, while the geometric mean of the GTFP index of the urban agglomeration in the middle reaches of the Yangtze River is the smallest.

Within the urban agglomeration, the GTFP index of the Yangtze River Delta urban agglomeration is relatively balanced, while the GTFP index of the core cities in the middle reaches of the Yangtze River and Chengdu and Chongqing urban agglomeration is obviously higher than that of the surrounding cities. Second, from the perspective of time trend, the GTFP index of the three major urban agglomerations has basically experienced an upward trend from 2006 to 2008 and a downward trend from 2008 to 2011. On the one hand, 2006 is the first year of China's economic green transformation, especially Wuhan city circle and Changsha-Zhuzhou-Xiangtan of Hunan Province in the middle reaches of the Yangtze River, which were designated as "two oriented society" comprehensive reform pilot zone by the state during this period; On the other hand, the international financial crisis in 2008 and the four trillion investment stimulus aiming at steady growth in 2009 brought downward fluctuation pressure on the quality of regional economic development.

Table 2. Green total factor productivity index of three urban agglomerations in Yangtze River Economic Belt

Year	2004	2005	2006	2007	2008	2009	2010	2011	2012	2013	2014	2015	2016	Geometric mean
Yangtze River Delta	1.025	0.979	1.069	1.057	1.040	1.063	1.045	1.014	1.061	0.997	1.011	1.046	1.125	1.040
Middle reaches of Yangtze River	0.963	0.942	0.996	1.049	1.105	1.050	1.024	1.018	1.081	1.001	1.021	1.36	1.126	1.031
Chengdu and Chongqing	1.006	0.926	1.042	1.051	1.075	1.053	1.040	1.055	1.086	0.993	1.062	1.20	1.033	1.033

4.3.2 Spatial Correlation Test of GTFP Index

Spatial correlation test is an important standard to judge whether to use traditional panel model or spatial econometric model. Table 3 reports the spatial correlation test results of green total factor productivity of Yangtze River Delta urban agglomeration, middle reaches of Yangtze River urban agglomeration and Chengdu and Chongqing urban agglomeration. It can be seen from Table 3: Firstly, Moran's I index based on geographical weight of the three major urban agglomerations is significantly positive at the significance level of 10%, indicating that there is a positive spatial correlation between urban GTFP index in urban agglomerations. Overall, the Moran's I index of Yangtze River Delta urban agglomeration is the largest, followed by Chengdu and Chongqing urban agglomeration, and the urban agglomeration in the middle reaches of Yangtze River is relatively small. Secondly, the Moran's I index of the Yangtze River Delta and Chengdu and Chongqing urban agglomerations shows an increasing trend year by year, indicating that the spatial dependence is gradually increasing, while the Moran's I index of the middle reaches of the Yangtze River urban agglomerations shows a trend of first increasing and then decreasing during 2004 to 2016. It shows that the gap of GTFP index of cities in the middle reaches of the Yangtze River urban agglomeration is expanding.

Table 3. Moran's I index of GTFP of three major urban groups in Yangtze River Economic Belt

Year	2004	2005	2006	2007	2008	2009	2010
Yangtze River Delta urban agglomeration	0.156^{***}	0.152^{***}	0.167^{**}	0.163^{**}	0.169^{**}	0.162^{***}	0.168^{**}
Middle reaches of Yangtze River urban agglomeration	0.063^{*}	0.061^{**}	0.074^{**}	0.086^{**}	0.083^{**}	0.092^{**}	0.099^{**}
Chengdu and Chongqing urban agglomeration	0.121^{**}	0.102^{**}	0.114^{**}	0.125^{**}	0.128^{**}	0.133^{***}	0.138^{**}
Year	2011	2012	2013	2014	2015	2016	
Yangtze River Delta urban agglomeration	0.172^{**}	0.176^{**}	0.181^{**}	0.18^{**}	0.183^{**}	0.189^{**}	
Middle reaches of Yangtze River urban agglomeration	0.121^{*}	0.115^{**}	0.107^{**}	0.103^{*}	0.099^{**}	0.094^{**}	
Chengdu and Chongqing urban agglomeration	0.147^{*}	0.141^{***}	0.146^{**}	0.153^{*}	0.158^{***}	0.162^{***}	

Note: *, **, ***, respectively indicate that they are significant at 10%, 5% and 1% levels

4.3.3 Test of the Impact of Producer Services Agglomeration on High Quality Development

In order to analyze the impact of different agglomeration modes of producer services industry on high-quality economic development of the three major urban agglomerations in the Yangtze River economic belt, this paper uses three spatial panel models of

SEM, SAR and SDM to estimate. The results are shown in Table 4. Considering the log likelihood value, modified R^2, LR and Wald, SDM model is better than SAR and SEM models. Therefore, SDM model is selected as the econometric model, and the corresponding fixed effect model is selected after Hausman test. The estimated results in Table 4 show that the estimated value of ρ or ϕ of the three urban agglomerations is significantly positive at the level of 1%, which indicates that there is spatial externality of the city GTFP index. The quality of economic development of a city in the urban agglomeration will be affected by the quality of economic development of other cities nearby. At the same time, the spatial spillover effect in urban agglomeration is not only reflected in the dependent variables, but also in the influence of core explanatory variables and control variables on the dependent variables. From the preliminary estimation results of the three spatial econometric models, we can see that the high-quality development level of urban economy in urban agglomeration is not only affected by the agglomeration mode of local producer services, but also affected by the spatial spillover effect of producer services industry agglomeration in other cities in urban agglomeration.

Table 4. Estimation results of three spatial econometric models

Statistics and coefficients	Yangtze River Delta urban agglomeration			Middle reaches of Yangtze River urban agglomeration			Chengdu and Chongqing urban agglomeration		
	SEM	SAR	SDM	SEM	SAR	SDM	SEM	SAR	SDM
lnSP	0.0887**	0.0892**	0.0942***	0.0826**	0.0814*	0.0927***	0.0805***	0.0695**	0.0912***
lnDV	0.0895*	0.0906**	0.0961***	0.0891*	0.0787**	0.0884***	0.0817**	0.0715**	0.0891***
lnHum	0.0815**	0.0772**	0.0625**	0.0783**	0.0695***	0.0658***	0.0619***	0.0643***	0.0551***
lnFdi	0.0485**	0.0507*	0.0413**	−0.0175*	0.0093	−0.0251**	0.0315**	0.0332**	0.0287**
lnInf	0.0347**	0.0339**	0.0323***	0.0497**	0.0471***	0.0485***	0.0413***	0.0407***	0.0385***
lnRoad	0.0214**	0.0196**	0.0107**	0.0272**	0.0314**	0.0182**	0.0276**	0.0315**	0.0127**
lnGov	0.1098***	0.1126**	0.1282***	0.0917	0.0886	0.1437***	0.0896**	0.0914*	0.1712***
lnWJ	0.0762**	0.0821**	0.0806***	0.0803***	0.0821**	0.0793***	0.0698***	0.0747***	0.0786***
W*lnSP	–	–	0.1095***	–	–	0.0130	–	–	0.1027***
W*lnDV	–	–	0.1283**	–	–	−0.0122**	–	–	0.0112
W*lnHum	–	–	0.0247**	–	–	0.0068	–	–	0.0095
W*lnFdi	–	–	0.0009	–	–	0.0032	–	–	0.0009
W*lnInf	–	–	0.0307***	–	–	0.0327***	–	–	0.0392***
W*lnRoad	–	–	0.0139*	–	–	0.0164**	–	–	0.0128**
W*lnGov	–	–	−0.1592*	–	–	−0.3016**	–	–	−0.3094**
W*lnWJ	–	–	0.0026*	–	–	−0.0081**	–	–	−0.0028*
ρor ø	0.5201***	0.4985***	0.5071***	0.4825***	0.4503***	0.4216***	0.4805***	0.4687***	0.4695***
LR	67.362***	69.781***	71.573***	61.749***	60.805***	62.389***	63.593***	64.694***	66.013**
L-Likelihood	1286.167	1479.583	1801.095	1240.848	1457.462	1782.957	1227.135	1474.536	1694.478
R^2	0.738	0.723	0.782	0.712	0.695	0.726	0.704	0.752	0.773

Note: *, **, ***, respectively indicate that they are significant at 10%, 5% and 1% levels

4.3.4 Analysis of Spatial Spillover Effect of Producer Services Agglomeration

In the above SDM model with global effect setting, the parameter estimates of independent variables do not directly reflect all the effects of explanatory variables on the explained variables, so the regression coefficients of explanatory variables in Table 4 can not be used to explain the marginal effects of each explanatory variable on high-quality economic development. Therefore, according to the method of Lacey and Pace (2009), this paper further uses the partial differential method of spatial regression model to decompose the spatial spillover effect of each explanatory variable into direct effect and indirect effect. Table 5 reports the direct, indirect and total effects of the specialization and diversification agglomeration of producer services in the three major urban agglomerations of the Yangtze River Economic Belt on high-quality economic development.

From the regression results of the Yangtze River Delta urban agglomeration, the regression coefficients of the specialization agglomeration and diversification agglomeration of producer services industry are all significantly positive, and the regression coefficient of the indirect effect of diversification agglomeration is the largest, which indicates that producer services industry agglomeration of the Yangtze River Delta urban agglomeration can not only promote the high-quality development of local economy, but also benefit the high-quality development of the adjacent cities in the urban agglomeration. As far as the industrial agglomeration mode is concerned, the diversified agglomeration of producer services industry in the Yangtze River Delta urban agglomeration is more conducive to high-quality economic development than the specialized agglomeration, whether in the direct or indirect effects. The possible reason is that the diversified agglomeration is more conducive to the spillover of complementary knowledge, skills, technology and other heterogeneous resources among different industries in the urban agglomeration network, and is conducive to the formation of a good ecological environment for industrial innovation and development in the industrial chain of producer services, so as to promote high-quality economic development. As far as the source of industrial agglomeration effect is concerned, the indirect effect of producer services industry agglomeration is greater than the direct effect. The high-quality economic development of Yangtze River Delta urban agglomeration benefits more from the "diffusion effect" of producer services industry agglomeration among cities, which means that producer services industry agglomeration has formed an effective mechanism of industrial benign interaction in urban agglomeration. The Yangtze River Delta urban agglomeration was the first to put forward the regional cooperation mechanism of "three-level operation, integration and division", which is not only the institutional cooperation on the economic level among cities, but also the institutional cooperation on the social, population, resource and environmental system level.

From the regression results of the urban agglomeration in the middle reaches of the Yangtze River, the regression coefficients of the two agglomeration models are significantly positive, and the direct effect of specialized agglomeration is greater, which shows that the specialized agglomeration and diversified agglomeration of producer services industry have a significant role in promoting the high-quality development of local economy, and more dependent on the specialized agglomeration of producer services

Table 5. Direct effect, indirect effect and total effect of SDM model of three urban agglomerations

Variable	Yangtze River Delta urban agglomeration			Middle reaches of Yangtze River urban agglomeration			Chengdu and Chongqing urban agglomeration		
	Direct effect	Indirect effect	Total effect	Direct effect	Indirect effect	Total effect	Direct effect	Indirect effect	Total effect
lnSP	0.0913*** (4.06)	0.1262*** (4.35)	0.2175*** (4.87)	0.1052*** (4.24)	0.0130 (1.53)	0.1182** (2.29)	0.0945*** (4.11)	0.1136*** (4.31)	0.2081*** (4.75)
lnDV	0.0982*** (4.12)	0.1307*** (4.42)	0.2289*** (4.91)	0.0904*** (3.92)	−0.0129** (−2.23)	0.0775*** (3.58)	0.0910*** (4.05)	0.0104 (1.51)	0.1017** (2.53)
lnHum	0.0612*** (3.83)	0.0204** (2.05)	0.0816*** (4.16)	0.0575*** (3.59)	0.0068 (0.83)	0.0643*** (3.89)	0.0551*** (3.56)	0.0105 (1.37)	0.0656*** (3.87)
lnFdi	0.0451*** (6.67)	0.0013 (1.29)	0.0464*** (6.97)	−0.0104*** (−3.67)	0.0032 (1.04)	−0.0072 (−1.26)	0.0287*** (4.73)	0.0012 (1.36)	0.0228*** (5.65)
lnInf	0.0321*** (4.02)	0.0318*** (3.92)	0.0639*** (4.87)	0.0409*** (4.35)	0.0327*** (4.03)	0.0736*** (4.98)	0.0385*** (4.22)	0.0309*** (3.86)	0.0694*** (4.93)
lnRoad	0.0135** (2.65)	0.0151* (1.72)	0.0286*** (3.08)	0.0157*** (2.91)	0.0164** (2.59)	0.0320** (3.23)	0.0127** (2.53)	0.0156** (2.53)	0.0383** (3.41)
lnGov	0.1602*** (5.07)	−0.1709* (−1.80)	−0.0107** (−2.52)	0.1513*** (4.97)	−0.3016** (−2.48)	−0.1503** (−4.42)	0.1712*** (5.54)	−0.3120** (−2.33)	−0.1408* (−2.07)
lnWJ	0.0815*** (3.68)	0.0012* (2.05)	0.0827*** (3.75)	0.0823*** (4.01)	−0.0081** (−2.14)	0.0742*** (3.52)	0.0786*** (3.61)	0.0031* (−1.79)	0.0755*** (3.23)

Note: *, **, ***, respectively indicate that they are significant at 10%, 5% and 1% levels; t statistic in brackets

industry. However, in terms of indirect effect, the coefficient of specialized agglomeration is not significant, while the coefficient of diversified agglomeration is significantly negative, indicating that the diversified agglomeration of producer services industry in the middle reaches of the Yangtze River urban agglomeration is not conducive to the high-quality economic development of surrounding cities, that is, there is a "siphon effect". The possible reason is that the intercity cooperation of the urban agglomeration in the middle reaches of the Yangtze River is still in the stage of "layout cooperation", and the adjustment of regional layout planning is the adjustment of production relations. However, the industrial development relationship between cities is more competitive, and the problem of industrial homogeneity still exists. There is still a big gap between the development goal of regional economic integration of the urban agglomeration. As far as the total effect is concerned, the total effect of producer services industry agglomeration in the middle reaches of the Yangtze River is significantly lower than that in the Yangtze River Delta. The main reason is that the positive spatial spillover channel of industrial agglomeration is blocked. This shows that the key to enhance the agglomeration effect of producer services industry in the middle reaches of the Yangtze River lies in strengthening the coordinated development of the industries in the urban agglomeration, abandoning the idea of development in its own way, accelerating the transition of urban cooperation from "layout cooperation" to "factor cooperation" and "system cooperation" with "overall thinking", and opening up the spillover channel of intercity indirect effect of producer services industry agglomeration, Accelerate the realization of industrial integration development of urban agglomeration in the real sense.

From the regression results of Chengdu and Chongqing urban agglomeration, the regression coefficients of direct effect and indirect effect of producer services specialization agglomeration are significantly positive, and the indirect effect is greater than the direct effect. The regression coefficient of direct effect of diversified agglomeration is significantly positive, but the indirect effect is not significant. On the one hand, the specialized agglomeration of producer services in Chengdu and Chongqing urban agglomeration can not only promote the high-quality development of the local economy, but also benefit the high-quality development of the neighboring cities in the urban agglomeration. The specialized agglomeration promotes the high-quality development of the urban agglomeration economy through economies of scale, labor pool and knowledge spillover within industries. On the other hand, the diversified agglomeration of producer services industry in Chengdu and Chongqing urban agglomeration only has the local effect of promoting high-quality economic development, but the inter city spillover effect has not yet appeared. The possible reason is that the diversified agglomeration effect is mainly produced by knowledge spillover among different industries, which is closely related to the integration development of producer services among different industries. The more perfect the diversified support platform for industrial integration development and the institutional system of industrial cooperation in different places, the more conducive to the formation of urban agglomeration innovation network, The higher the demand of the surrounding cities for their productive services, the more effective the inter city spatial spillover effect of diversified agglomeration can be. The intercity cooperation of Chengdu and Chongqing urban agglomeration is mainly manifested in the regional cooperation at the level of production factors. Therefore, we should speed up the regional

cooperation mode from the stage of factor cooperation to the stage of institutional cooperation, and improve the platform construction of diversified cross regional development
of producer services industry at the institutional level.

The estimated results of the control variables are as follows. First, the direct effect of
human capital on the high-quality economic development of the three urban agglomerations is significantly positive, while the indirect effect is only significantly positive in the
Yangtze River Delta urban agglomerations, indicating that human capital reserve plays
an important role in the high-quality economic development of urban agglomerations.
Compared with the Yangtze River Delta urban agglomeration, the inter city spillover
effect of human capital in the middle reaches of the Yangtze River and Chengdu and
Chongqing urban agglomeration is not significant, so it is necessary to strengthen the
system and mechanism construction of talent flow and talent introduction. Second, in the
Yangtze River Delta and Chengdu and Chongqing urban agglomeration, the direct effect
of foreign capital participation on high-quality economic development is significantly
positive, while the indirect effect is not significant. In the urban agglomeration of the
middle reaches of the Yangtze River, the direct effect of foreign investment participation on high-quality economic development is significantly negative, and there may be a
pollution haven effect of FDI. Third, the direct, indirect and total effects of information
infrastructure on the high-quality economic development of the three urban agglomerations are all significantly positive, which indicates that information construction, as
the infrastructure for the transformation of traditional manufacturing industry and the
upgrading of modern service industry, is conducive to promoting the cross city knowledge and technology dissemination and rapid transaction of producer services industry
through information technology, so as to improve the quality of economic development of urban agglomeration. Fourth, as the cornerstone of urban agglomeration system
construction, the regression coefficients of direct and indirect effects of transportation
infrastructure are significantly positive, but less than the marginal effect of information
infrastructure. Fifth, the direct effect of government intervention on high-quality economic development is significantly positive, while the indirect effect and total effect
are significantly negative. The possible reason is that local protectionism makes it difficult for different cities to reach a consensus in the formulation of industrial policies,
energy conservation and emission reduction standards and joint prevention and control
of environmental pollution, and there may be self-interest strategies such as competition of production factors and environmental dumping among local governments. Sixth,
the direct and total effects of environmental regulation on the high-quality economic
development of the three urban agglomerations in the Yangtze River economic belt are
significantly positive, but the indirect effects are significantly negative in the middle
reaches of the Yangtze River and Chengdu and Chongqing urban agglomerations. This
shows that a strong level of environmental regulation can effectively promote the high-
quality development of local economy, but the urban environmental regulation of the
middle reaches of the Yangtze River and Chengdu and Chongqing urban agglomeration
is not conducive to the high-quality economic development of neighboring cities. The
possible reasons are the lack of institutional cooperation at the ecological and environmental protection level of the urban agglomeration, and the potential difference of

environmental regulation level among cities promotes the gradient transfer of polluting enterprises in the urban agglomeration.

5 Research Conclusions and Policy Recommendations

The results show that producer services industry agglomeration can promote the high-quality economic development of the three urban agglomerations in the Yangtze River economic belt, but there are obvious differences in the sources of industrial agglomeration effect among the three urban agglomerations.

First, the direct and indirect effects of the specialization and diversification of producer services agglomeration on high-quality economic development are significantly positive, and the inter city effect of diversification agglomeration is an important source of producer services agglomeration effect. In other words, the agglomeration of producer services, especially the diversified agglomeration, can not only promote the high-quality development of local economy, but also benefit the high-quality development of neighboring cities in the Yangtze River Delta urban agglomeration. The agglomeration of producer services has obvious "diffusion effect" in the Yangtze River Delta urban agglomeration.

Second, the direct effects of specialization and diversification of producer services agglomeration are positive, but the indirect effects of specialization agglomeration are not significant, while the indirect effects of diversification agglomeration are negative. This means that the local effect is the reason why producer services industry agglomeration promotes the high-quality economic development of the urban agglomeration in the middle reaches of the Yangtze River. Producer services agglomeration in the urban agglomeration in the middle reaches of the Yangtze River is only conducive to the high-quality development of the local economy, but it has a "siphon effect" on the economic development of other cities in the urban agglomeration.

Thirdly, the direct and indirect effects of producer services specialization agglomeration in Chengdu and Chongqing urban agglomeration are positive, the direct effect of diversification agglomeration is positive, but the indirect effect is not obvious. In other words, the specialized agglomeration of producer services in Chengdu and Chongqing urban agglomeration can not only promote the high-quality development of local economy, but also benefit the high-quality development of neighboring cities in the urban agglomeration. However, the inter city spillover effect of diversified agglomeration of producer services has not yet appeared. Fourth, the estimation of control variables shows that the high-quality development of urban economy in the urban agglomeration is not only related to the level of local human capital, the level of informatization, the intensity of environmental regulation, the degree of government intervention and transportation infrastructure, but also related to the relevant factors of other neighboring cities in the urban agglomeration.

Based on the above research conclusions, the following three policy recommendations are made:

First, relying on the two agglomeration modes of specialization and diversification, accelerate the agglomeration development of producer services industry in the three major urban agglomerations of the Yangtze River economic belt. We should fully understand the role of producer services industry agglomeration in improving the high-quality economic development of urban agglomeration, combine with the industrial structure and manufacturing foundation of urban agglomeration, scientifically position the function of producer services, and choose a reasonable and effective producer services agglomeration mode. As far as the Yangtze River Delta urban agglomeration is concerned, the economic aggregate and technical level are the "leading" of the Yangtze River economic belt, and the industrial development is relatively mature. We should ensure the specialization scale of high-end producer services, and pay more attention to the diversified agglomeration of producer services industry. As far as the urban agglomerations in the middle reaches of the Yangtze River and Chengdu and Chongqing urban agglomerations are concerned, we should focus on building specialized producer services agglomeration areas at this stage, and form high-quality producer services specialized agglomeration by cultivating producer services matching the development needs of local manufacturing industry, so that the diversified agglomeration of producer services can be built on the basis of higher quality specialized agglomeration.

Second, promote the spatial gradient development and network layout of producer services industry in urban agglomerations to open up space for the inter city spillover effect of producer services industry agglomeration. It is necessary to strengthen the relevance and complementarity of producer services industry in urban agglomerations, form a reasonable gradient and cooperation network of industrial development among cities, and build a modern industrial system of urban agglomerations with the integration and symbiosis of producer services and industry and efficiency orientation. First of all, strengthen the division and cooperation of producer services industry among cities, and form an industrial pattern with reasonable functional division, distinctive features and complementary advantages between core cities and surrounding small and medium-sized cities. On the one hand, give full play to the radiation driving function of the agglomeration effect of producer services industry in core cities such as Shanghai, Nanjing, Hangzhou, Wuhan, Chongqing and Chengdu; On the other hand, the surrounding small and medium-sized cities should also make full use of the high-end productive service talents and technical resources of the core cities to transform and upgrade their own manufacturing industry, and accelerate the transformation from the traditional extensive production style with high energy consumption and high pollution to the green and low-carbon high value-added production style. Secondly, we should build an innovation chain focused on the industrial chain and transform the potential energy of science and technology into innovation kinetic energy. Relying on the new "core edge" structure of "producer services industry—manufacturing" in urban agglomeration, we should build a complete industrial chain of R & D innovation, processing and manufacturing, and supporting services in urban agglomeration, form the input-output relationship and main and auxiliary relationship of urban agglomeration economy in urban agglomeration, and let different micro subjects obtain positive externality of knowledge spillover through learning mechanism.

Third, strengthen the construction of inter city cooperation mechanism at the institutional level of urban agglomeration, and promote the economic integration development of urban agglomeration. The empirical study shows that the difference of agglomeration effect of producer services industry in the three urban agglomerations of the Yangtze

River economic belt is mainly reflected in whether it can play a positive inter city effect. Therefore, local governments should not only consider the spatial accessibility between cities, but also strengthen the long-term mechanism of exchange and sharing of human capital and information. First of all, on the premise of innovating regional integration policies and starting with the spatial layout of strategic service facilities in urban agglomerations, we should promote the transformation of regional cooperation mode from layout cooperation, element cooperation to institutional cooperation. On the basis of promoting free competition, local governments should improve the cooperation mechanism, strive to build an open urban network across geographical boundaries, and gradually eliminate market segmentation and factor circulation restrictions among cities. Secondly, it is necessary to establish a "talent-platform-system" three in one inter city industrial cooperation support system to provide policy basic support for high-quality industrial development. Through the establishment of cross regional coordination agencies, we can promote the flow of innovation elements across regions and industries, and give full play to the dual role of local effect and inter city effect of producer services industry agglomeration. In addition, we should further optimize the environmental regulation system of the Yangtze River economic belt, improve the transfer mechanism of industrial agglomeration, and prevent the gradient transfer of polluting industries.

References

Sun, B., Ding, S.: Is big city beneficial to the economic growth of small city: evidence from the Yangtze River Delta urban agglomeration. Geograph. Res. 35(9), 1615–1625 (2016)

Mei, Z., Xu, S., Ou, Y., et al.: Spatiotemporal evolution of urban spatial interaction in Pearl River Delta Urban Agglomeration in recent 20 years. Geoscience 32(6), 694–701 (2012)

Ke, S., Xia, J.: Agglomeration effect and backflow effect of Central Plains urban agglomeration. China Soft Sci. (10), 93–103 (2010)

Sun, D., Zhang, J., Hu, Y., et al.: Analysis of the formation mechanism of "metropolitan shadow area" based on industrial spatial connection – a comparative study of Yangtze River Delta urban agglomeration and Beijing Tianjin Hebei urban agglomeration. Geograph. Sci. 33(9), 1043–1050 (2013)

Wang, T., Zeng, J.: Spatial pattern analysis of intercity competition and cooperation of urban agglomeration in the middle reaches of the Yangtze River – based on the perspective of urban competitiveness and spatial interaction. Trop. Geograp. 34(3), 390–398 (2014)

Zhang, X., Chen, S., Liao, C., et al.: Spatial spillover effect of economic development in Beijing Tianjin Hebei region. Geograph. Res. 35(9), 1753–1766 (2016)

Yu, Y., Gao, X., Wei, P.: The impact of producer services agglomeration on the high quality economic development of urban agglomerations: a case study of the three major urban agglomerations in the Yangtze River economic belt. Urban Probl. (7), 56–66 (2020)

Chen, A., Partridge, M.D.: When are cities engines of growth in China? Spread and backwash effects across the urban hierarchy. Reg. Stud. 47(8), 1313–1331 (2013)

Wang, F., Xie, J.: A study on the growth rate of green total factor productivity of provinces in China. China Popul. Sci. (2), 53–62 (2015)

Li, B., Qi, Y., Li, Q.: Fiscal decentralization, FDI and green total factor productivity: an empirical test based on panel data dynamic GMM method. Int. Trade Issues (7), 119–129 (2016)

Zhang, J., Tan, W.: Study on the green total factor productivity in main cities of China. Proc. Rijeka Sch. Econ. 34(1), 215–234 (2016)

Liu, H., Li, C., Peng, Y.: Research on the regional gap and regional collaborative improvement of green total factor productivity in China. China Popul. Sci. (4), 32–43 (2018)

Zheng, Q.: The impact of urbanization on green total factor productivity – an analysis based on the threshold effect of public expenditure. Urban Issues (3), 48–56 (2018)

Li, M., Wang, J.: The quality of foreign direct investment and the growth of green total factor productivity in China. Soft Sci. **33**(9), 13–20 (2019)

Jin, B.: Economic research on "high quality development". China's Ind. Econ. (4), 5–18 (2018)

Ma, R., Luo, H., Wang, H., Wang, T.: Research on evaluation index system and measurement of high quality development of regional economy in China. China Soft Sci. (7), 60–67 (2019)

Liu, S., Zhang, S., Zhu, H.: Measurement of national innovation driving force and its effect on high-quality economic development. Res. Quant. Econ. Technol. Econ. (4), 3–23 (2019)

Zhang, Y., Dong, Q., Ni min.: Differentiation between the development of service industry and "structural slowdown" –also on building a modern economic system with high quality development. Econ. Trends (2), 23–35 (2018)

Li, H., Shi, J.F.: Energy efficiency analysis on Chinese industrial sectors: an improved super - SBM model with undesirable outputs. J. Clean. Prod. (65), 97–107 (2014)

Shan, H.: Re estimation of China's capital stock K: 1952–2006. Res. Quant. Econ. Technol. Econ. (10), 17–31 (2008)

Hall, R.E., Jones, C.I.: Why do some countries produce so much more output per worker than others? Q. J. Econ. **144**(1), 83–116 (1999)

Tu, Z.: Coordination of environment, resources and industrial growth. Econ. Res. (2), 93–105 (2008)

Duranton, G., Puga, D.: Diversity and specialization in cities: why, where and when does it matter? Urban Stud. (3), 533–555 (1999)

Ke, S., Ming, H., Yuan, C.: Synergy and co - agglomeration of producer services and manufacturing: a panel data analysis of Chinese cities. Reg. Stud. **48**(11), 1829–1841 (2014)

Xi, Q., Chen, X., Li, G.: Research on the mode selection of China's Urban Producer Services – guided by the improvement of industrial efficiency. China's Ind. Econ. (2), 18–30 (2015)

Keller, W.: Geographic localization of international technology diffusion. Am. Econ. Rev. **92**(1), 120–142 (2002)

Elhorst, J.P.: Applied spatial econometrics: raising the bar. Spat. Econ. Anal. **5**(1), 9–28 (2010)

LeSage, J., Pace, R.K.: Introduction to Spatial Econometrics, pp. 34–48. CRC Press, Taylor & Francis Group (2009)

Liu, Y., Xia, J., Li, Y.: Agglomeration of producer services and upgrading of manufacturing industry. China's Ind. Econ. **35**(7), 24 (2017)

Limits and Transcendence of Global Governance and Service

Lindong Zhao[1] and Cunling Liu[2(✉)]

[1] Shenzhen Personnel and Talents Public Service Center, Guangdong 5181040,
People's Republic of China
[2] Shenzhen Institute of Information Technology, Guangdong 518172,
People's Republic of China

Abstract. Global problems such as environmental problem, security problem and food problem need global governance and service. But global governance and service is western governance. Coin has its two sides: on the one hand, under the background of globalization, Global problem needs global governance and service; On the other hand, Global governance reflects developed country's values, which can be used as interference tool in domestic affairs. The two sides make global governance and service has utopian limitations. Only beyond its own limitations, global governance and service will become real political.

Keywords: Globalization · Global governance and service · Global problem

1 Introduction

1.1 The Emergence of Global Governance and Service and Global Problems

From the perspective of globalization, global governance is a response to global problems. Globalization, especially economic globalization makes the economy influence beyond the national borders restrictions. Rapid globalization strengthen the interdependence between countries, under the influence of rapid globalization, most of the countries enhanced their contact; all sorts of problems associated with each other; domestic politics is closely associated with international politics. Rapid economic growth in China, India, Brazil, Russia and other emerging countries makes the interdependence of the global economy reach a new height. Economic globalization promotes the internationalization of the political. Global problems have emerged under the background of globalization, which is a major driving force of global governance. "So-called global problem, is refers to the contemporary international community faced a series of transcending national and regional boundaries, in relation to the entire human survival and the development of serious problems." [1] Global problems have the following characteristics: first, the influence of universality. The influence of the global issues has beyond the limits of national boundaries, which has broad implications for all countries in the world. Second, the complicated problem. Under the background of globalization, global problem is the result of combined action of many factors. Third, interdependence. Global problem

© Springer Nature Switzerland AG 2022
M. A. Serhani and L.-J. Zhang (Eds.): SERVICES 2021, LNCS 12996, pp. 53–60, 2022.
https://doi.org/10.1007/978-3-030-96585-3_4

calls for the joint efforts of all countries in the world and need to seek a systemic solution. Fourth, benefit and harm. Due to the unequal political status, the economic crisis and ecological crisis has been transferred from the developed countries to developing countries. The developed countries have got the interests and developing countries often have to bear the consequences of environmental damage and shortage of resources. Fifth, Value differences. Global issues reflect different values and outlook on development. Today, a series of global problems such as north and south, war and peace, ecological imbalance, food crisis, resources shortage, population, refugees, drugs, AIDS, international human rights and nationalism, terrorism have caused widespread attention from the international society. No matter from the scale, scope, or the consequences, these problems is global problems and their solutions have global significance. The nature of global problems determines what they need is not unilateral but multilateral joint action, not unilateral individual decision-making but a global public policy, planning and comprehensive treatment [2].

1.2   The Hegemony of Crisis and the Global Political Awakening

From the perspective of developed countries, the global governance is the response to global political awakening and the crisis of western hegemony. After World War II, a new upsurge of national independence and national liberation movement has emerged. Nationalist movement and the campaign against colonial rule eventually have lead to the collapse of the old colonial rule. Independent of the third world countries has impact the capitalist world system. After the end of the cold war, the European Union, Japan and other regional organizations have affected the hegemony of the United States. The rises of emerging countries, such as China, Russia, India, Brazil, etc. are changing the distribution of power in the world political. The pattern of the world political develops towards multi-polarization, for the only superpower, the multi-polarization is a serious threat to the United States hegemony. Faced with crisis of hegemony and global political awakening, us-led western countries need to look for a new tool to dominate the world situation. The global political democratization and the inner hidden inequality of the global governance satisfy this need. Global governance of the real question is: whose governance? Who is involved in? Who's leading? Who set the rules? These problems exposed the internal conflict of global governance. Due to the awakening of the global political and it's relatively weakened strength, the United States need to find new ways to maintain the dominance of world politics. For us-led western countries, global governance has become an important means to promote their international strategy. At the practice level, global governance indicates the change of the intervention strategy. Global democratization has become a kind of rhetoric, which has the effect of modification and camouflage.

1.3   Democratic Wave of the World Promote

From the perspective of developing countries, global governance Suggested that developing countries demand for a new international political order. The democratic wave appeared on the one hand is influenced by western "exporting democracy", on the other hand also reflects that the developing countries demand for democratic political. In the specific historical stage, developing countries face dual task: the political construction and the economic construction. The fundamental purpose of Learning western lies in the rapid development. The developed countries focus on the maneuverability of global governance, but developing countries emphasize the equality of global governance. As the change of power distribution, some developing countries scholars advocated "power in global politics should be redistributed between developed countries and developing countries". The attraction of global governance lies in the good political desire. "Governance concept has attracted the extensive attention of scholars, which mainly because with the advent of the era of globalization, the human political life is undergoing major transformation. One of the most striking changes is the center of gravity of the human political process has changed: from ruling to governance, from good politics to good governance, from government to governance without government, from the national government to global governance" [3].

2   The Limits of Global Governance and Service

The limitations of global governance and service are mainly manifested in the following aspects:

2.1   The Limits of the Governance Body

"Global governance and service body mainly has three categories: (1), governments, government departments and the national government authorities; (2), the formal international organization, such as the United N ations, the world bank, world trade organization, the international monetary fund, etc.; (3), informal civil society organizations around the world." [4] The governance body has its limitation. From the perspective of governance entity, global governance is different from the domestic governance. The government is the governance entity in domestic governance. But global governance has no clear governance entity. "World country" or "world government" does not exist. Due to the popular sovereignty, the pursuit of different interests and the deep concern about the effectiveness of the current governance mechanisms, Global governance cannot construct a world government. Even if the idea of world government is possible, it will meet with opposition because of weakening the existing national sovereignty. Given the limitations of itself, the United Nations is also difficult to become a explicit governance entity. To this, it was pointed out that the United Nations can not develop into the "world government", a similar domestic government can't appear on the earth. [5] From the perspective of government capacity, state sovereignty is not the same as decision-making ability. Some African governments can not perform their decision-making in real. Developing countries facing the double pressure: domestic political stability and economic development. The developed countries also have different decision-making ability and influence. At the international level, because of the different national power, the influence of the gap between governments is more obvious. From the Angle of informal global civil society, the international organizations, multinational corporations, non-governmental organizations in global governance are mostly from the developed countries. It is difficult to guarantee the independence of its own activities. The process of global governance is vulnerable to manipulation of the developed countries. Non-state actors in developing countries are not perfect, which lead to weak in the governance mechanism.

2.2   The Limits of Governance Objectives

Governance objectives reflect the multiple conflicts of interest between the governance bodies.

Most emerging countries committed to the international and regional peace and stability in order to create better conditions for their economy development. Us-led developed countries committed to maintaining its dominant position in world politics.Global governance target reflected the position of developed countries. Us-led western countries promote global governance in the name of human rights and democracy. Human rights have become the excuse for intervention in the internal affairs. America sees himself as the measure of democracy. The purpose of pursuing democratic values is to enhance the attraction and influence of the United States. Some scholars believe that "the reality of the international system is not anarchy. Since the second world war, especially since the end of the cold war, the United States is in the center of the world order" [6]. Dominant countries claim to maintain the status quo. When the dominance is shaken, they will promote the status change in the favorable direction. A major manifestation is trying to establish a new international organization to replace the United Nations. Developing countries is playing a more and more important effect in the United Nations. There are differences between the developed countries and developing countries in global governance. Us-led western countries advocated more subjectivity governance. They want to set up a new coalition of organizations outside the United Nations. In global governance, they want to promote non-state organization participation. These actors mainly include NGO, regional organizations, civil society and multinationals. Because non-state actors mainly come from the developed countries, the diversity of the governance body conforms to the desire of the developed countries.Most of the developing countries are more likely to traditional management. They advocate the government as the center in global governance.

2.3   The Limitation of the Governance Rules

At present, global governance rules is not perfect. Existing governance rules are set by the developed countries, which reflect the western values and interests. Developing countries is at the periphery of the governance framework. They have no say in the formulation of global governance rules. Governance rules formulated by the developed countries serve the interests of developed countries. The existing governance rules have no mandatory and binding. There are three difficulties in global governance: the contradiction between the sovereign state and the international system; the contradiction between national interests and global interests; a conflict between countries with the global rules. [7] Multiple governance body is not equal to the equality participation. The reality of national interests and the fiction of global interests lead to disputes in the process of rule-making. But global regulation will limit state sovereignty. The legitimacy and the rationality of the global regulation have been questioned by the sovereign state. The legitimacy of governance rules depends on the equal participation and universal respect for the interests of the countries. Due to the unequal rights and obligations, the legitimacy of governance rules is hard to come by in the international political environment. Global governance is trying to develop a new regulation to manage the domestic and international affairs. The fact shows that global governance is still far away from the ideal goals.

2.4 The Limits of Governance Values

About the value of global governance, global governance committee believes that: "to improve the quality of global governance, we need global citizen moral to guide our global action; we need a kind of moral leadership. We called for core values, including respect for life, liberty, justice and fairness, mutual respect, love and integrity." [8] The developed countries and developing countries have different development concept. The developed countries think democracy promotes economic development. Developing countries think that economic growth will promote the democratization process. Under the background of globalization, developing countries and developed countries face different tasks. In a specific historical coordinate system, developing countries need to develop national economy and maintain social stability. Its domestic problems are more important than international issues. Their domestic problem is an important part of international problems. Developed countries are faced with the problem of maintaining and strengthening its dominant position, when their dominance is impacted by the emerging power. They need to create international environment that conducive to the interests of the country. As you can see, in the context of global governance, the behavior of the developed countries is extravert. The behavior of developing countries is introverted. Extraversion behavior may boost international cooperation; it could also mean intervention in the interior of the country. Introverted behavior may maintain their country's sovereignty independence; it may also lead to alienation.

2.5 The Limits of Governance Results

Global governance has not obtained ideal result. There are apparent conflicts between the global general interest and the embodied national interest. Nationalism, national self-interest limited joint action in the process of global governance. Under the background of globalization, the intended competition has become the obstacle of the global cooperation. Hegemonism and national self-interest help with each other. Developed countries through the hegemony to achieve their own purposes; they also use national self-interest to interfere in internal affairs. Realism paradigm outweighed the idealism of the paradigm. The military balance, political and diplomatic soft balance has become important nation strategy. Neoliberal economist's plan was met with serious questions. "New liberalism in the 70s is the product of economic liberalization. The new liberalism was imposed on other countries, eventually evolved into the financial crisis." PASCAL, Lord, criticized all sorts of false theories: "since the new liberalism theory pursued in nearly 30 years, those various popular theories once dominated the ideological and theoretical which introduced by the influential people actually are unfounded. It is a kind of fraud." [8] David carter criticized some blind confidence: "in 2002, economists are confident economics has developed to the point of such a" science. "The government will have new tools and knowledge to prevent great depression." [10] This is enough to justify a single economic thinking can't cope with the complexity of global issues.

3 The Possibility of Global Governance and Service Beyond Its Limitations

Whether to beyond its limits depends on several factors.

3.1   Interest Game and World Order

World political reality still is based on national power competition. World order maintenance has the following pattern: (1), power balance mode, by retaining power balance to maintain the world order. (2), the hegemony pattern, safeguard the order of the world by a hegemony country. (3), the bipolar pattern, through both sides of the checks and balances to maintain order in the world. (4), multipolar political system, maintain the world order by the interaction between multiple political bodies.For the weak, the order from the strong is a kind of imposing order. For the strong, the order from the weak is not binding. The focus of the problem is world order will serve its builder. Realism believes that there is struggle for power and wealth between countries. Due to the power and wealth finiteness, the contradiction between countries needed to be solved through the way of war. Global governance offers a new framework to solve global problems. However, global governance is a normative and empirical area. Global governance process depends fundamentally on the possibility of consensus interest based on game. Only beyond its limitations, global governance may break through idealistic political barriers and become a reality.

3.2   Sovereign Governance or Human Rights Governance

Human rights, democracy and freedom are very controversial topics for scholars both at home and abroad. To avoid these theoretical debates, in terms of international political reality, there are two points. One point is "want human rights, but not international intervention", "want democracy, but not global democratization". Some scholars believe that in the process of global governance, sovereign state faces the following difficulties: one is the contradiction between integral of global public problem and the exclusive of sovereignty; Second, the strengthening of international organization and the weak of national sovereignty; Third, the rise of global civil society and the erosion of national sovereignty independence. [10] They point out that in the process of global governance, if the sovereign state resists global cooperation on the grounds of safeguarding state sovereignty, it is hard for global governance to get a wide range of support. It is difficult to play their role. In global governance practice, therefore, how to play the leading role of a sovereign state is critical.

3.3   One Party Leading or Multi-participation

To beyond its own limits, global governance needs to think constructing a participation mechanism.

If dominated by one country, the essence of the global governance is hegemony governance. If dominated by the west, global governance has become the western governance. Governed by non-state actors also does not represent the interests of developing countries. One party dominant cannot solve global problems. Henry Kissinger thinks that multiple crisis and turmoil in the world may give the Obama administration opportunity to rebuild the world order. However, as the only dominant world power, the United States is unable to independently solve a series of global problems produced by its own policies and actions. [11] One party dominant cannot bring legitimacy for global governance. Unilateral hegemony governance faced with challenges. Global governance should promote multi-participation rather than one party dominant.

3.4 Power Distribution and Responsibility Sharing

In the established global governance framework, there are unequal distribution of power and responsibility sharing. Different national power determines their different voice. For liability share problem, there are differences between developed countries and developing countries. The cost and the income sharing in the global governance are not equal. About global problems, developed countries and developing countries are at different historical stages. The sustainable development solution given by the developed countries is not necessarily suitable for requirements of developing countries. Developed countries have entered the stage of "late capitalism". On one hand, the developed countries imports energy, resources and primary goods from developing countries. On the other hand, they control exports of advanced technology strictly. Developing countries faced with more and more pressure from environment and the social development. But they have no enough technology and capital to realize the transformation of economic structure. Under the background of globalization, the gap between the rich and the poor is bigger and bigger. This further reduces the possibility of matching the powers and duties. The result will be that the later development pays for the problem produced by the previous development. Accrual allocation determines the equality of global governance.

3.5 Western Values and Pluralistic Values

"God gave every people a cup, people drink into their life through the cup." The result of globalization is the cup has been broken. A lot of national culture and value has been hit. Globalization did not the transcendent the north-south gap; the north-south gap is wide. Under the influence of cultural globalization; each behavior body in global system has changed. But this does not mean that effects are the same. Globalization will certainly bring the consciousness of humanity. On the contrary, however, this process did not bring cultural homogeneity. It has caused more cultural differences and the expansion of local culture. To date, a global value system does not exist. Global governance should respect diverse value and the western values should not be promoted as a universal value. The west needs to rediscover its social culture. Cultural diversity decided the different people can choose a different path.

4   Conclusion

The crisis and the challenge of globalization determine the global governance and service is a topic that can not be avoided. The universality of global issues suggests that global governance is necessary; the concreteness of the state interests suggests that global governance is complicated. Only beyond its limitations, global governance may break through idealistic political barriers and become a political reality.

References

1. Cai, T.: Global Problems and Contemporary International Relations, p. 2. Tian Jin People's Publishing House, Tianjin (2002). (in Chinese)
2. Chen, C.: Global governance research in China. Cass J. Polit. Sci. **25**(1), 118–126 (2009). (in Chinese)
3. Yu, K.: An introduction to global governance. Marxism Reality **13**(1), 20–32 (2002). (in Chinese)
4. Yu, K.: An introduction to global governance. Marxism Reality **13**(1), 20–32 (2002). (in Chinese)
5. Xie, X.: Global governance problems. Hu-xiang BBS **32**(2), 120–122+128 (2009). (in Chinese)
6. Pang, Z.: Hegemony governance and global governance. Foreign Comments **26**(4), 16–20 (2009). (in Chinese)
7. Liu, X.: The concept or reality: contradiction in the theory of global governance. J. Jilin Univ. Soc. Proc. **4**, 85–90 (2008). (in Chinese)
8. Zhou, Y.: Focus on the future trends and thinking development strategy of China. Foreign Theory **22**(1), 1–7+72 (2012). (in Chinese)
9. Citrin, J.: The government trust. Foreign Theory **22**(10), 29–42 (2012). (in Chinese)
10. Li, H.: Sovereign state participates in global governance problem under the background of globalization. Bus. Inf. **3**(4), 73+86 (2008). (in Chinese)
11. Pang, Z.: Hegemony governance and global governance. Foreign Comments **26**(4), 16–20 (2009). (in Chinese)

An English to Urdu Educational Video Translation Pipeline to Reinforce Mother-Tongue Based Learning

Navid Shaghaghi[1]([⊠]) [iD], Smita Ghosh[1], Fatima Ali[1], and Abdul Basit Ali[2]

[1] Ethical, Pragmatic, and Intelligent Computing (EPIC) Laboratory and the Frugal Innovation Hub (FIH), Departments of Computer Science and Engineering (CSE), Mathematics and Computer Science (MCS), and Information Science and Analytic (ISA), Santa Clara University, Santa Clara, CA, USA
{nshaghaghi,sghosh3,fali}@scu.edu
[2] Department of Computer Science, San Jose State University, San Jose, CA, USA
abdulbasit.ali@sjsu.edu
https://www.scu.edu, https://www.sjsu.edu

Abstract. As the world's technological capacity to store information grew, Information Communication Technology (ICT) fused into almost every aspect of our lives and became one of the most important priorities in the field of education. Furthermore, educational technology is often not available to most or is in a sub-par state of functioning, especially in the more rural areas of many countries. As the recent global pandemic has further exacerbated this inequality, education has had to shift to online delivery of content via videos (live or prerecorded), audiobooks, and electronic text. Over the years considerable research exploring the role of language and utilization of visual aids in education has shown that visual learning in one's native language, is significantly more effective and yields higher retention rates. This is challenging for students in third-world countries as quality educational material in their language are often lacking. Inspired by the above reasons, this paper proposes a video translation pipeline from English to Urdu as a service, which goes beyond simple translation and incorporates the cultural needs of the learners to reinforce learning in a multilingual education system.

Keywords: Education Technology (EdTech) · Humanitarian Information Communications Technology (ICT) · Language translation · Natural Language Processing (NLP) · Bridging digital divide · Quality education

1 Motivation

Due to the 2020–21 COVID-19 global pandemic, the education of students worldwide has experienced a near two-year gap! At the onset of the pandemic, students in developed countries were quickly transitioned onto online learning platforms

© Springer Nature Switzerland AG 2022
M. A. Serhani and L.-J. Zhang (Eds.): SERVICES 2021, LNCS 12996, pp. 61–74, 2022.
https://doi.org/10.1007/978-3-030-96585-3_5

and lectures resumed shortly after via the numerous video conferencing platforms available such as Zoom Meetings, Microsoft Teams, Cisco Webex Business, Google Meet, etc. Students in developing countries, however, were not so
lucky. Some of them were eventually able to get access due to wealth or certain local governments stepping in to make resources available quickly, but most
were simply sent home to wait till schools reopened. As a result, most students
in developing countries such as Pakistan [1,2,26] and India [17–19], have fallen
behind on their education.

To help students in these areas, lecture videos and other online material
used for educating students in the US can for instance be used to fill the gap.
Lecture videos are very effective as they provide a visual component to learning
which cannot be achieved by audiobooks or simply reading textbooks on various
subjects. However, these videos and material are all in English [37] while the
official language of Pakistan and its neighboring parts of India is Urdu [13].
Hence, if these videos are to be used at all, they need to first be translated.

Translation is a difficult and time-consuming task that can not be done effectively without many hours of labor. Even translating one subject material for
one class grade is almost an impossible undertaking for a single individual. Furthermore, any translation that will be successful in educating the students must
be more than a mere literal translation of the words from English to Urdu. Sur-
passing all of the grammatical differences between the two languages, the cultural contexts of the words and sentences need to be considered. For instance, if
an example makes American pop culture references in order to connect with
students, those examples will most likely be meaningless to students in the
remote villages of Pakistan. Hence, the task of translating such educational
videos becomes even more cumbersome and daunting enough to scare away many
who would like to help.

In an effort to help Urdu speaking students around the world, Santa Clara
University's Ethical, Pragmatic, and Intelligent Computing (EPIC) research lab-
oratory and Frugal Innovation Hub (FIH) in California, USA have partnered with
ClassRoute Innovation Ventures Pvt. Ltd. in India to automate the translation of
educational material from English to various languages [33,34] including Urdu.
The contributions of which are:

1. A cost-effective and efficient translation pipeline that enhances mother-
   tongue-based learning
2. A translation system that incorporates the cultural needs of the learners to
   reinforce learning in a multilingual education system
3. A means of quality education for students in countries where English is not
   a primary language
4. The incorporation of neuro-linguistic programming and education-based
   research into an online educational platform

This paper reports the developed pipeline for semi-automatic transcription,
translation, and reproduction of educational videos and material from English
to Urdu and delineates the in-progress and future improvements for further
automating the pipeline. Section 2 describes the effectiveness of the focus on

developing Mother-Tongue (Urdu) based visual learning educational material (videos) in achieving the goal set out for this project. Section 3 describes the pipeline proposed in this paper and provides a description of the different tools experimented with, along with each of their strengths and weaknesses. Section 4 reports the initial results for the tools explored for each step of the pipeline and provides a discussion of the results. Sections 5 and 6 describe the currently under development and future steps for the project respectively. Lastly, Sect. 7 provides some concluding remarks.

2 The Efficiency of Visual and Mother-Tongue Based Education

The way to make this effort most effective for the most number of students is to utilize visual learning techniques in the learner's mother tongue.

2.1 Visual Learning

Visual learning has been proven to help improve the period for which information is retained by learners. Visual learning focuses on combining videos and images to leave a lasting memory and is directly processed by long-term memory [4,31]. Pedagogical research into teaching videos has been broadly positive and focused on their benefits for students' experience as well as how specific features of videos can enhance learning and retention [14]. In [15] the authors analyzed the use of video as a mediating artifact on an interpretive approach framed as authentic participant-centered inquiry and employed multiple theoretical frameworks to generate perspectives on the affordance and constraints of learning from videos. Countless research explores audio-visual tools (like pictures, animation videos, and films) in modern curriculum [3] and has shown that they play a significant role in quality education. They serve as a motivational tool in enhancing students' attention in reading literary texts [32].

But why focus on visual learning only when there are different types of (auditory, visual, verbal, etc.) learners? According to [5] Verbal learners are a group that constitutes about 30% of the general population who learn by hearing. They benefit from class lectures and from discussion of class materials in study groups or in oral presentations, but chafe at written assignments. Experiential learners are about 5% of the population and they learn by doing and touching, and hence clinical work, role-playing exercises, moot courts, and the such are their best instructional modalities. However, visual learners constitute the remaining 65% of the population which constitute a majority. Visual learners need to see what they are learning, and while they have difficulty following oral lectures they perform well at written assignments and readily recall material they have read [5]. Thus, Visual Learning is the most effective learning method when trying to target the largest segment of a population.

2.2   Learning in Mother-Tongue

Language plays an important role in quality education. For instance, in [29] the authors explore the power of using native languages in courses and concluded that language contributes to making teaching and learning mathematics more inclusive. Quality education should be delivered in the language spoken at home. When learners use their home language to learn rather than another language, their understanding and performance is likely to improve. Being able to move between two languages lessens the cognitive load (the brain having to do too many tasks at once) and enables learners to better explain what they know and can do [6]. However, this minimum standard is not met for hundreds of millions of learners, hence limiting their ability to develop foundations for learning. A study cited by the Global Education Monitoring Report (GEMR) and published by UNESCO found that as much as 40% of the global population does not have access to education in a language they speak or fully understand [24]. Learning in a regional language improves the rate at which one learns and helps with retention rates. Minimal access further marginalize[s] people - usually indigenous people, women, children, older people, and persons with disabilities - who don't speak the official or dominant language(s) in a given region [23]. This lack of education, caused by language barriers, contributes to increasing poverty rates, especially in third-world countries. The solution to which is increased access to education in regional languages.

Understanding content is easier and faster in one's native language, as it does not require as much external research to comprehend. The GEMR also agrees with this stance as they found that using the home language as the language of instruction has a positive impact on learning across the board. They emphasize that it is crucial for children to build a strong foundation in critical skills such as literacy and numeracy. To accomplish this task effectively, schools need to teach the curriculum in the language which children understand best (their mother tongue). Researchers have also found that doing so directly correlates with improved performance in the second language and other subjects [24] as speaking one's mother tongue in school increases self-confidence and thinking skills, and conveys freedom of speech [25]. Inspired by these studies, for the past decade the world has seen increasingly resolute and genuine interest in mother tongue-based multilingual education (MTB-MLE), a broad framework of educational provision promoting the use of learners' first languages or mother tongues as the primary media of instruction. MTB-MLE appears mainly in two broad political contexts of education: the first in educating different linguistic minority groups found in a particular country which nonetheless deploys a foreign language or a national language as the main medium of instruction, and the second in using the mother tongues in 'mainstream' education, supplanting erstwhile languages of education [36].

3   Pipeline Description

The developed pipeline takes as input a video with audio in one language and outputs a translated video with audio in another language. English Educational videos produced for remote education during the COVID-19 pandemic were selected as input and the Urdu language which is spoken predominantly in Pakistan and parts of India was selected as the target language to translate the videos into. In future iterations, the translation pipeline will also incorporate cultural contexts in order to better connect the educational material covered in the videos with the learners. Figure 1 shows this flow. The pipeline is further broken down into the following steps:

1. Transcription: English audio is converted to corresponding English text.
2. Grammar Checking: the grammar of the transcript obtained from the transcription step is checked and improved upon.
3. Translation: the English text is translated to Urdu text.
4. Audio Generation: an Urdu voice-over is generated for the original video.

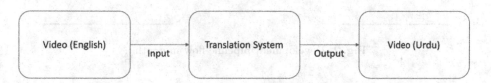

Fig. 1. Development pipeline flow overview

Figure 2 shows the flow of the intermediate steps. As depicted, the transcription and translation steps require multiple actions. The actions associated with these steps are:

1. Finding an existing software or API that performs the task
2. Using the software or API to test and find the accuracy
3. Perfecting the transcript
4. Logging changes in order to change or enhance the output to account for correctness and (in the near future) cultural context.

While experimenting with step one it was determined that there could be a possibility of incorrect grammar in the audio, resulting in a transcript with incorrect grammar. So, following step one, a Grammar checking internal step was added between transcription and translation. This step revolves around fixing the grammar of the transcript obtained from step one before translation is attempted. The final step involves generating a voice-over using the new transcript. This step is composed of the two intermediate steps of:

1. Recording the new audio in-sync with the video
2. Merging the new audio with the video

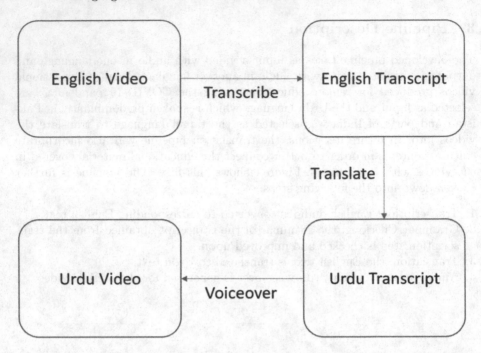

Fig. 2. Development pipeline flow in-depth

4   Results and Discussion

With advancement in Natural Language Processing (NLP), an abundance of tools exist for transcription from English audio to English text as well as the translation of English Text to Urdu text.

4.1   Transcription APIs

Many APIs and tools were experimented with and the accuracy of each tool was documented in Table 1. As can be observed, the tools all have comparable high accuracies. An error rate was nevertheless calculated by measuring the ratio of the number of incorrect and missing words generated by the transcription tool to the total number of words in the original transcript. The accuracy was then determined by subtracting the above value from 1 and multiplying it by 100. The formula for accuracy calculation is given in Fig. 3.

$$\left(1 - \frac{number\ of\ incorrect\ words\ and\ missing\ words}{number\ of\ words\ in\ original\ transcript}\right) * 100$$

Fig. 3. Formula to calculate accuracy for transcription

Table 1. Transcription programs and their accuracies

API	Accuracy (%)	Cost
Otter.ai	93–95	$99.96/year, $12.99/month (PRO)
Google Cloud	90–95	$0.006/15 s
IBM Watson	85–88	$0.01/minutes (PLUS)
Audext	85–88	$30/month
Temi	80–83	$0.25 per audio minute

Otter.ai. The first software experimented with was Otter.ai [30]. Otter is an artificial intelligence-based API that translates English audio and video to English text. Initial results show that Otter has an accuracy of 95%. Otter transcribes the audio to text almost word for word and does so at an efficient pace. After the transcription is complete, the transcribed text is annotatable via the platform. This makes it easy to note any mistakes with the transcription and keep track of them. The transcription can also be exported through a text file, and the platform makes it easy to share and collaborate with groups. It is trusted and used by several organizations such as Verizon, Zoom, Dropbox, IBM, Columbia University, UCLA, RBC Royal bank, Paloalto Networks etc.

However, Otter.ai does have its drawbacks. For example, it incorrectly detects inflections in voices and reads them as the end of a sentence or a question, creating a punctuation problem. Also, if the speaker is not clear with their diction, it can cause Otter to translate the audio literally, which causes inconsistencies in the transcription. Figure 4 is an example of the actual transcript versus the transcript generated by Otter.ai. The errors have been highlighted in yellow.

Google Speech-to-Text. The second API that was looked at for transcription was Google's cloud-based speech-to-text software [11]. The main advantages of this software are that the transcript acquired is very accurate (similar to Otter.ai) and it offers an infusion feature that allows the user to infuse the speech transcription into any application with the API itself. Furthermore, there are different domain-specific models which one can choose from to transcribe the video: default, command/search, phone call, and video. This helps best optimize the transcription output based on the input. One of the primary use cases for this software is multimedia transcription.

The main functional problem found with this software is that the auto-punctuation feature does not always work as intended. For example, it can mistake the speaker in the video pausing mid-sentence as a period. This is the only but major flaw with this software that was observed.

IBM Watson, Audext, and Temi. The other three APIs experimented with were IBM Watson [16], Audext [20] and Temi [35]. They had similar grammar and punctuation issues, but these problems occurred more often and hence they

This ball is lying at rest on the ground. It's stationary; not moving at all. Is there any force acting on it? Listen to my question carefully. Are there any forces acting on the ball when it is stationary? If your answer was a no or if you didn't have an answer, it means you don't know the concept of force yet. The answer is yes; there are forces acting on this ball. There is a gravitational force which is trying to pull the ball towards the centre of the earth. And the ground is applying an equal force exactly in the opposite direction. This force is called the normal force. Because these two forces are balanced, they do not change the position of an object. The net force acting on the ball is zero. So what does this tell you? Just humans pushing or pulling is not the only kind of force. And just because there are forces acting on an object will not mean the object will move. For the object to move, there has to be some net force.

This ball is lying at rest on the ground. It's stationary, not moving at all. Is there any force acting on it? Listen to my question carefully. Are there any forces acting on the ball when it is stationary? If your answer was no, or if you didn't have an answer, it means you don't know the concept of force yet? The answer is yes, there are forces acting on this ball. There is a gravitational force, which is trying to pull the ball towards the center of the earth. And the ground is applying an equal force exactly in the opposite direction. This force is called the normal force. Because these two forces are balanced, they do not change the position of an object the net force acting on the ball is zero. So what does this tell you? Just humans pushing up pulling is not the only kind of force. And just because there are forces acting on an object will not mean the object will move. For the object to move, there has to be some net force. So now let's say a person

Fig. 4. Original transcript vs. Otter.ai-generated transcript (Color figure online)

This ball is lying addressed on the ground. It's stationary not moving at all. Is there any force acting on it? Listen to my question carefully. Are there any forces acting on the ball when it is stationary? If your answer was a no, or if you didn't have an answer, it means you don't know the concept of force yet. The answer is yes, there are forces acting on this ball. There is a gravitational force, which is trying to pull the ball towards the center of the earth. And the ground is applying an equal force exactly. In the opposite direction. This force is called the normal force because these two forces are balanced. They do not change the position of an object. The net force acting on the ball is zero. So what does this tell you? Just humans pushing up pulling is not the only kind of force. And just because there are forces acting on an object will not mean the object will move for the object to move. There has to be some net force.

Fig. 5. Temi-generated transcript

had lower accuracy. Figure 5 is an example of the transcript generated by Temi (lowest accuracy, as can be seen by the numerous highlights).

Based on the accuracy results obtained, Otter.ai is the best tool for the transcription step as it fits the needs best. It was very accurate (95% accuracy) and it did not have the punctuation issue that Google's software did.

4.2   Translation

The first software experimented with for this step was Google Translate [10] which was found to be 65–70% accurate at translating English text to Urdu Text. Accuracy in this context has a very specific definition: it is defined as the accuracy of the transcript within the context of the subject being discussed. Effort was made to gauge the accuracy in relation to how native Urdu-speaking instructors would teach the given topic.

It was observed that there were some drawbacks to using this API:

1. It translates the sentence literally, in the sense that it is a word-for-word translation.
2. The translation does not always reflect typical, everyday dialogue as it may be difficult for even native speakers to understand, especially within the context of the subject(s) they are trying to learn.

Figure 6 shows an example of the original transcript in English versus the translated version in Urdu where the following inaccuracies were observed:

1. In the sentence "If your answer was a no, or if you didn't have an answer, it means you don't know the concept of force yet", the English word "no" was translated as the Urdu word "Aik" which means "one" in English. Furthermore, the incorrect version of the English word "of" was used. The correct equivalent Urdu word should be "ka", not "ko".
2. In the sentence "Because these two forces are balanced, they do not change the position of an object", the English word "position" was translated as the Urdu word "haisiyat" which means "position", but within the context of a social/political situation and not in the context of Math or Science. The correct word would be "Jagah".
3. The sentence "Just humans pushing or pulling is not the only kind of force" is translated to "Sirf insaan ko dhaka dena ya khichna hi taakat ki ek kisam nahi hai", Which means that "Not only pushing or pulling people is the only kind of force." It should have instead been "Sirf insaan ka dhaka dena ya khichna hi taakat ki ek kisam nahi hai".
4. In the sentence "And just because there are forces acting on an object will not mean the object will move", the English word "object" was translated as the Urdu word "aitraaz" which means "objection".
5. In the sentence "For the object to move, there has to be some net force", the English word for "net" was translated as the Urdu word "khalis" which means "fine".

Due to the low level of accuracy, the translation step is a semi-automatic step. Meaning the English text obtained from the transcription step is fed into the translator but the Urdu text returned by the translator needs to be verified and perfected by a native Urdu speaker.

Fig. 6. English vs. Urdu translation comparison

4.3   Audio Generation

The third and final step in the pipeline is to record a voice-over of the original video using the Urdu text. One software explored for automating this step was murf.ai [22]. This software is very accurate for English voice-overs and mapping to videos, however, it does not work for Urdu.

Furthermore, regardless of the language, it is often difficult to comprehend what AI-generated voices are saying, even to native speakers. Hence, even if a successful API is found, it remains to be tested for effectiveness in the context of multilingual education.

Given the lack of software and APIs to automate the Urdu voice-overs, the Urdu audio needs to be manually recorded and spliced into the video in place of the original English audio. Interestingly, Fiverr [8], a freelancer hiring web service already exists which includes native speaker freelancers available for hire for voice-overs. A quick inspection of the site returns many posts from freelancers about their willingness to offer English to Urdu voice-over services at a reasonable cost [9].

5   Work in Progress

5.1   Further Automation of the Translation Step of the Pipeline

Google translate is not the only translation API even though it is the most used. Other translation APIs are being explored for accuracy and if any have higher accuracy then they will be slotted in place of the Google Translate API. However, it is very likely that better APIs are not publicly available. Therefore, the results of the translation step are being kept and tracked in order to analyze the drawbacks as well as to incorporate those into the design of a custom NLP model that can provide a secondary translation step where the API's result is sent in as input and a more accurate translation is outputted.

6 Future Work

6.1 Addition of Cultural Context to the Translation Step of the Pipeline

As observed, even though Google translate had an accuracy of 70%, it did have the issue that the translation does not always reflect typical, everyday dialogue that may be difficult for native speakers to understand, especially within the context of the subject(s) they are trying to learn. Research work has been done in designing English to Urdu translation models using Machine Learning and Natural Language Processing [21,27,28]. More existing translation models from English to Urdu will be explored and analyzed before designing a new translation model that would not only accurately perform the translation but also incorporate the cultural context with respect to a multilingual education system. One option yet to be explored is Google's cloud-based machine learning software: AutoML [12]. AutoML makes it easy to train machine learning models using the data set one provides. Currently, a dataset with five hundred sentences in English is being translated to Urdu. After creating the English to Urdu data set, it will be used to train and validate the new translation model.

6.2 Incorporation into ClassRoute [7]

Once the pipeline is fully developed, it will be incorporated into *ClassRoute*. *ClassRoute* is a multi-language educational community learning platform and content generator which aims to disseminate quality educational content to those who have restricted access to education. The internet is filled with forums and platforms where people can virtually gather to socialize and learn. *ClassRoute* is unique as its online learning community caters to the specific needs of students. Currently, the educational content is all in text format but inspired by the efficiency and effectiveness of visual-based learning in one's mother tongue, this platform wants to incorporate a video feature that makes quality educational videos available to all its users.

7 Conclusion

With the heightened importance of visual-based learning in mother-tongue for improving the quality of education received by students in developing countries, such as Pakistan and India, that have been left without proper schooling due to the global COVID-19 pandemic, the necessity of an automated (or even semi-automatic) pipeline for translating existing English content into other languages such as Urdu is Paramount! After experimenting with existing tools for each of the steps of the pipeline, it was observed that certain tools performed really accurately for some steps. For example, Otter transcribes video content with a 95% accuracy. But more exploration is needed for tools that can be used in the translation step. So far, Google translate, translates English to Urdu with a 70%

accuracy but does not reflect the typical, everyday dialogue which is difficult for native speakers to understand, especially within the context of the subject(s) they are trying to learn.

Research will continue for automating the final audio generation and video voice-over generation step as currently it is being done manually. As the goal of the project is to reinforce learning in a mother tongue-based language, research will be continued in perfecting the translation by incorporating cultural context in order to support learning in a native language.

Acknowledgements. A very special thanks is due to Shamsa Ali, an Urdu native speaker and mother to authors Fatima and Abdul Basit Ali, for personally verifying the very many Urdu translations made by the system. And to Abhishek Biswas, founder and CEO of ClassRoute Innovation Ventures Pvt. Ltd. who inspired the project through his and his company's efforts to address the educational gap faced by disenfranchised students in India and neighbouring countries. And lastly, to the department of Mathematics & Computer Science (MCS) of the College of Arts & Sciences (CAS), department of Computer Science and Engineering (CSE) and the Frugal Innovation Hub (FIH) of the School of Engineering (SoE), and the department of Information Systems & Analytics (ISA) of the School of Business at Santa Clara University, as well as, to the department of Computer Science at San Jose State University for all their continued support of the project.

References

1. Anwar, M., Khan, A., Sultan, K.: The barriers and challenges faced by students in online education during COVID-19 pandemic in Pakistan. Gomal Univ. J. Res. **36**(1), 52–62 (2020)
2. Aslam, S., Parveen, K.: The challenges of online teaching in COVID-19 pandemic: a case study of public universities in Karachi, Pakistan (2021)
3. Badalova, B.: Effectiveness of audio-visual aids in teaching process. Acad. Res. Educ. Sci. **2**(4), 1905–1909 (2021)
4. Bautista-Vallejo, J.M., Hernández-Carrera, R.M., Moreno-Rodriguez, R., Lopez-Bastias, J.L.: Improvement of memory and motivation in language learning in primary education through the interactive digital whiteboard (IDW): the future in a post-pandemic period. Sustainability **12**(19), 8109 (2020)
5. Bradford, W.C.: Reaching the visual learner: teaching property through art. The Law Teacher, vol. 11 (2004)
6. British Council: Seven reasons for teachers to welcome home languages in education (2018). https://www.britishcouncil.org/voices-magazine/reasons-for-teachers-to-prioritise-home-languages-in-education
7. ClassRoute Innovation Ventures Pvt. Ltd.: Quality education for all (2021). http://classroute.org
8. Fiverr (2010). https://www.fiverr.com
9. Fiverr (2010). https://www.fiverr.com/search/gigs?query=voice+over+in+urdu
10. Google: Google translate (2006). https://translate.google.com/
11. Google: Google speech-to-text (2013). https://cloud.google.com/speech-to-text
12. Google: Google cloud automl (2018). https://cloud.google.com/automl
13. Gujjar, A.A.: Third language through first language-shifting the paradigm. i-Manager's J. Sch. Educ. Technol. **3**(2), 49 (2007)

14. Harrison, T.: How distance education students perceive the impact of teaching videos on their learning. Open Learn. J. Open Distance e-Learn. **35**(3), 260–276 (2020)
15. Higgins, J., Moeed, A., Eden, R.: Video as a mediating artefact of science learning: cogenerated views of what helps students learn from watching video. Asia-Pac. Sci. Educ. **4**(1), 1–19 (2018). https://doi.org/10.1186/s41029-018-0022-7
16. IBM: IBM Watson speech-to-text. https://www.ibm.com/cloud/watson-speech-to-text
17. Jadhav, V.R., Bagul, T.D., Aswale, S.R.: COVID-19 era: students' role to look at problems in education system during lockdown issues in Maharashtra, India. Int. J. Res. Rev. **7**(5), 328–331 (2020)
18. Jena, P.K.: Impact of pandemic COVID-19 on education in India. Int. J. Curr. Res. (IJCR) **12**(7), 12582–12586 (2020)
19. Kapasia, N., et al.: Impact of lockdown on learning status of undergraduate and postgraduate students during COVID-19 pandemic in West Bengal, India. Child Youth Serv. Rev. **116**, 105194 (2020)
20. Kate: Audext. https://audext.com/
21. Khan, S.N., Usman, I.: A model for English to Urdu and Hindi machine translation system using translation rules and artificial neural network. Int. Arab J. Inf. Technol. **16**(1), 125–131 (2019)
22. Murf.ai. https://murf.ai/
23. Öktem, A., Jaam, M.A., DeLuca, E., Tang, G.: Gamayun-language technology for humanitarian response. In: 2020 IEEE Global Humanitarian Technology Conference (GHTC), pp. 1–4. IEEE (2020)
24. OXFAM India: If you don't understand, how can you learn? (2020). https://www.oxfamindia.org/blog/how-can-indias-education-system-escape-vicious-cycle-inequality-and-discrimination
25. Ozfidan, B.: Right of knowing and using mother tongue: a mixed method study. Engl. Lang. Teach. **10**(12), 15–23 (2017)
26. Qamar, T.Q., Bawany, N.Z., et al.: Impact of COVID-19 on higher education in Pakistan: an exploratory study. IJERI: Int. J. Educ. Res. Innov. (15), 503–518 (2021)
27. Rai, S.I., Khan, M.U., Anwar, M.W.: English to Urdu: optimizing sequence learning in neural machine translation. In: 2020 3rd International Conference on Computing, Mathematics and Engineering Technologies (iCoMET), pp. 1–6. IEEE (2020)
28. Rauf, S.A., Abida, S., Zahra, S., Parvez, D., Bashir, J., et al.: On the exploration of English to Urdu machine translation. In: Proceedings of the 1st Joint Workshop on Spoken Language Technologies for Under-Resourced Languages (SLTU) and Collaboration and Computing for Under-Resourced Languages (CCURL), pp. 285–293 (2020)
29. Robertson, S.-A., Graven, M.: Language as an including or excluding factor in mathematics teaching and learning. Math. Educ. Res. J. **32**(1), 77–101 (2019). https://doi.org/10.1007/s13394-019-00302-0
30. Liang, S., Fu, Y.: Otter.ai (2016). https://otter.ai
31. Schneider, S., Nebel, S., Beege, M., Rey, G.D.: The retrieval-enhancing effects of decorative pictures as memory cues in multimedia learning videos and subsequent performance tests. J. Educ. Psychol. **112**(6), 1111 (2020)
32. Shabiralyani, G., Hasan, K.S., Hamad, N., Iqbal, N.: Impact of visual aids in enhancing the learning process case research: district Dera Ghazi Khan. J. Educ. Pract. **6**(19), 226–233 (2015)

33. Shaghaghi, N., Ghosh, S., Kapoor, R.: Classroute: an English to Punjabi educational video translation pipeline for supporting Punjabi mother-tongue education. In: 2021 IEEE Global Humanitarian Technology Conference (GHTC), pp. 342–348 (2021). https://doi.org/10.1109/GHTC53159.2021.9612485

34. Shaghaghi, N., Israel, M.J., Ghosh, S., Mondal, S., Biswas, A.M.A.B.: Classroute: bridging the digital academic-content divide. In: 2021 1st Conference on Online Teaching for Mobile Education (OT4ME), pp. 168–173 (2021). https://doi.org/10.1109/OT4ME53559.2021.9638822

35. Temi (2017). https://www.temi.com/

36. Tupas, R.: Inequalities of multilingualism: challenges to mother tongue-based multilingual education. Lang. Educ. **29**(2), 112–124 (2015)

37. Van der Zee, T., Admiraal, W., Paas, F., Saab, N., Giesbers, B.: Effects of subtitles, complexity, and language proficiency on learning from online education videos. J. Media Psychol. **29**, 18–30 (2017). https://doi.org/10.1027/1864-1105/a000208

2021–2025: Forecast Analysis of Increased Housing Demand in Shenzhen

Xuan Li[✉] and Jinhong Xu

ShenZhen Institute of Information Technology, Shenzhen 518029, Guangdong, China

Abstract. High housing price and insufficient housing supply are the prominent problems facing the first-tier cities in China. And the affected factors of urban housing demand are so much, the stable housing market is so important in the process of urban development. This paper establishes a housing demand model to simulate and forecast the increased housing demand of residents in Shenzhen from 2021 to 2025 based on the housing, population, economy and human settlement survey over the years. The study found that, there will increase the demand of 1.2 million housing sets in the next five years, the current imbalance between housing supply and demand will still exist obviously in Shenzhen. So we need to standardize the development of housing rental market, vigorously develop public housing, in order to effectively alleviate the contradiction between supply and demand.

Keywords: Housing demand · Increased demand · Forecast analysis

1 Introduction

The housing demand is closely related to the economic development of the city. With the rapidly development of the city, the housing price is rising rapidly in the big cities. The realization of the housing demand of urban residents has become a research issue of social concern. Nowadays, the implementation of the three-child policy, the positioning of housing and housing not speculation, and the parallel promotion of the policies of both rental and purchase, the urban housing demand is changing. Whether the housing demand can be effectively satisfied has become an important factor for the city to attract and retain talents, and it is crucial for the healthy development of the city's economy and society [1, 2].

Over the past 40 years of reform and opening up, the builders of the Special Economic Zone have gathered in Shenzhen from all over the world and helped Shenzhen rise rapidly into a fully functioning megacities with a population of more than 20 million. Thanks to years of unremitting efforts, historic achievements have been made in housing, and the housing shortage has been basically solved. By the end of 2019, the total housing volume of the city was about 621 million square meters, with 11.29 million

Fund Project: ShenZhen Institute of information technology, University-level project of 11400-2021-010201-0126.

M. A. Serhani and L.-J. Zhang (Eds.): SERVICES 2021, LNCS 12996, pp. 75–82, 2022.
https://doi.org/10.1007/978-3-030-96585-3_6

units (rooms), basically meeting the housing demand of the city's more than 20 million management population. However, with the continuous net inflow of population, there will be more housing demand, which will put forward higher requirements for the housing development in Shenzhen.

Urban housing demand analysis is a hot issue in real estate economics. Mankiw and Weil (1989) studied the relationship between demographic structure transformation in the United States, housing demand and housing price, found that population could well explain the housing market changes in the past 20 years. Green and Hendershott (1996) studied the impact of population, income and education level on housing demand. Cui Yuying (2013) believes that housing demand is mainly influenced by economic factors, family factors, public resources factors and policy factors. Chen Yanbin and Chen Xiaoliang (2013) designed 9 schemes to estimate the scale of housing demand from three aspects, urbanization, family size change and population aging. Wu et al. (2016) measured the scale of new housing demand in 8 typical cities in China. Wu Jing and Xu Mandi (2021) designed a systematic method to quantitatively measure the scale of new urban housing demand based on census data, and measured and analyzed the scale of new housing demand in each region from 2001 to 2015 [3, 4].

On the basis of the above analysis, this paper analyzes the demand for new housing in the next five years based on the situation of population, housing, economy and other aspects of Shenzhen, in order to provide reference for Shenzhen to formulate housing policies and real estate enterprises to formulate development strategies (Table 1) .

Table 1. Change of housing floor area per capita in Shenzhen from 2015 to 2019

Year	Residential floor area (100 million square meters)	Per capita floor area of housing based on the managed population	Per capita floor space of housing based on permanent population
2015	5.71	30.65	50.18
2016	5.9	29.72	49.54
2017	6.07	28.45	48.45
2018	6.12	27.83	46.98
2019	6.21	23.88	46.2

2   Model Construction and Data Sources

2.1   Model Construction

Urban housing demand means that households have the ability to obtain and are willing to choose housing under the influence of certain income, price and other factors that affect housing demand. The housing demand can be divided into rigid demand, improvement demand, induced demand, and housing demand and rental demand. Factors affecting the total housing demand in a certain period include housing demand caused by urban population growth, improvement housing demand, inventory housing renewal demand, etc. In addition, the market also needs a certain proportion of vacant housing to maintain the normal circulation of housing sources [5, 6].

Based on the actual situation of Shenzhen's population development trend, economic development level, housing situation, residents' living behavior research and so on, the new housing demand of Shenzhen in the next five years is scientifically estimated. The new housing demand is divided into three parts: first, the housing demand of the new population; Second, the housing improvement demand of the existing population; Third, the demand caused by the renewal of stock housing. In addition, the market also needs a certain percentage of vacant housing to maintain the normal flow of housing supply. Therefore, the total new housing demand can be expressed as follows:

$$D_g = (D_P + D_s + D_c)/(1 - V) \tag{1}$$

Where, D_g is the total new housing demand, D_P is the housing demand caused by population growth, D_s is the housing demand for improvement, D_c is the housing demand caused by the renewal of stock housing, and V is the reasonable housing vacancy rate.

In the specific calculation, various components of housing demand in Formula (1) should be predicted by combining basic data and information such as building census data, housing registration data, population census, statistical yearbook and human settlement survey. D_P and D_s can be combined and calculated based on the analysis of the overall living level and population scale at the end of the planning period. The arrangement of the 14th Five-Year Urban Renewal Plan can be predicted comprehensively based on the historical data of the demolition and reconstruction scale of residential types during the 12th and 13th Five-Year Plan period and the latest urban renewal plan. The total new housing demand can be expressed as follows:

$$D_g = (A_2 \times (M_2 - M_1) + (A_2 - A_1) \times M_1 + D_c)/(1 - V) \tag{2}$$

Where, A_2 is the per capita housing floor area of the whole population in 2025, A_1 is the per capita housing floor area of the whole population in 2020, M_2 is the population in 2025, M_1 is the population in 2020, D_c is the housing demand caused by the renewal of housing stock, and V is the reasonable housing vacancy rate.

2.2 Data Sources

The basic information and data sources of housing demand forecast are relatively diverse, mainly including building census data, housing registration data, population census and management data, statistical yearbook data, human settlements sample survey data, stock housing renewal plan, etc.

2.2.1 Building Census Data

Building census is a detailed survey of housing stock, mainly including building total, number of floors, area, state, structure, floor area ratio, use and other attributes. Based on the building census data from 2015 to 2018 in Shenzhen, important information about housing quantity, type and structure was obtained in this study.

2.2.2 Housing Registration Data

Housing registration data contains more detailed information about housing on the list, which is ideal for analyzing housing stock. In this study, the relevant basic data were obtained from the grid management department and Shenzhen Real Estate Information Network.

2.2.3 Population Census and Management Data

Population censuses and 1% population sample surveys conducted by statistical departments provide housing information based partly on households, such as the number of rooms, the size of houses, amenities, etc. The statistical survey of managed population, grid population and introduced talents carried out by the Public Security Bureau, Grid Office and Human Resources and Social Security Bureau provides the relevant data of total urban population and new population structure with full caliber. In this study, basic information about population, structure and housing conditions was obtained from the Bureau of Statistics, Public Security Bureau, Grid Office and Human Resources and Social Security Bureau [7].

2.2.4 Statistical Yearbook Data

The data of urban population, per capita housing floor area and income structure in that year will be provided in the city statistical yearbook. Among them, the per capita housing floor area is obtained based on sampling survey, which may be greatly deviated from the actual situation. In this study, urban population, per capita housing floor area, income structure and other relevant data were obtained from the national and Shenzhen Statistical Yearbook.

2.2.5 Habitat Sampling Survey

The housing demand needs not only the full caliber housing data, but also needs to analyze the behavior characteristics of the living subject. Since 2006, Shenzhen has continued to conduct annual human settlements survey, which mainly focuses on the level of human settlements, housing status, satisfaction, willingness and other special surveys. Based on the statistical results of the human settlements survey, this study obtained the housing ownership rate, human living area, housing cost of commercial housing residents, housing choice, commuting time, vacancy rate and other relevant data in Shenzhen.

2.2.6 Urban Renewal and Shanty Reconstruction Plan

The intensity of land development and utilization in Shenzhen is high, and it is difficult to add new land. The redevelopment and utilization of existing land is an important way. The renovation of stock housing in Shenzhen mainly focuses on urban renewal, and there are a small number of renovation projects in shanty areas. Based on the data of Shenzhen's urban renewal and shanty reconstruction policies, project application and construction plan, this study analyzes the potential arising demands during the 14th Five-Year Plan period.

In general, data sources from different channels have their own advantages and disadvantages, and their statistical calibers may vary. In specific measurement, the "base" of housing should be determined by mutual checking and complementarity. For the key information that cannot be provided by all kinds of data, data exchange, special investigation, sampling survey, department interview and other methods are needed to supplement [8].

3 Analysis of New Housing Demand in Shenzhen

3.1 Population Forecast

In the process of city planning, mainly there are 9 kinds of methods to forecast the urban population scale, respectively is: the comprehensive method of growth rate, crop coefficient method, surplus labor transfer method, the demand for Labour method, regression analysis, the GM (1, 1) grey model method, elasticity coefficient method, the city level of economy scale and parkman's law. In this paper, the economic elasticity coefficient method is used to forecast and analyze the population scale of Shenzhen from 2021 to 2025.

Basic principle of forecasting method: calculate the growth rate of both population and GDP according to the number of years in the past, and the economic elasticity coefficient is the quotient of GDP growth rate and population growth rate.

$$P_n = P_0(1 + v')^n \tag{3}$$

$$v' = \frac{V'}{K} \tag{4}$$

$$P_n = P_0[(1 + v')] \tag{5}$$

In the formula, P_n is the size of urban population in the planning period, P_0 is the size of urban population in the base period, v' is the average annual population growth rate in the planning period, N is the planning period, v' is the average annual economic growth rate in the planning period, and K is the population elasticity coefficient of economic growth.

Forecasting steps, first, collect population and GDP data for at least five consecutive years in the same period and determine the average annual growth rate of both; Secondly, K value is determined according to the annual average growth rate of population and GDP. Thirdly, determine the annual average economic growth rate within the planning period, and get the annual average population growth rate within the planning period; Fourthly, the value is substituted in according to the above formula to predict the size of urban population in the planning period.

Based on the permanent population and GDP data of Shenzhen from 2001 to 2020, the scale of permanent population in 2025 is predicted. From 2001 to 2009, the growth rate of permanent resident population in Shenzhen reached 4%, and the GDP growth rate exceeded 15%. Population and economy were in the stage of rapid development at the same time. From 2010 to 2014, the growth rate of permanent resident population in Shenzhen slowed down, with an average annual growth rate of less than 1%. Meanwhile, GDP growth has slowed to around 10%. Since 2015, the population growth rate has accelerated, while the GDP growth rate is still declining, and the two are in reverse motion. This may be related to the loose household registration policy and talent introduction policy in our city in recent years. From 2016 to 2020, about 606,000 talents with bachelor degree or above will be introduced. According to the data provided by the Bureau of Human Resources, it is estimated that about 354,000 talents with bachelor degree or above will be introduced from 2018 to 2020.

In order to make the data smoothing better, the average annual growth rate of population and GDP in the past five years were calculated. It can be seen that the average annual growth rate of the permanent resident population in Shenzhen in the past five years fluctuated, with an average annual growth rate of 0.97% from 2010 to 2014, which was in a slow growth stage, and then the population growth rate became faster, with an average annual growth rate of 4.85% from 2014 to 2020. Over the same period, economic growth has slowed, averaging 13.24 percent from 2010 to 2014 and 10.16 percent from 2014 to 2020. In view of the inverse change between the annual population growth rate and the annual GDP growth rate in Shenzhen in recent years, in order to make the data prediction more accurate, the average annual population growth rate from 2009 to 2020 is 3.04%, and the average annual GDP growth rate in the same period is 12.36%. According to the economic development trend of Shenzhen, the average annual growth rate of GDP in 2021–2025 will be 5%–9%. In the case that Shenzhen's annual GDP growth rate is 5%, 7% and 9% respectively, the number of permanent residents in Shenzhen in 2025 is calculated.

If the average annual GDP growth rate is 5% from 2021 to 2025, the average population growth rate in the same period is 1.22%, then the urban population size will be about 14.18 million in 2025. When the GDP growth rate from 2021 to 2025 is 7%, the average rate of population growth in the same period is 1.71%, then the urban population size in 2025 is about 14.66 million. When the GDP growth rate from 2021 to 2025 is 9%, the average rate of population growth in the same period is 2.21%, then the urban population size in 2025 is about 15.18 million. Therefore, the population size of Shenzhen in 2025 will be between 14.18 million and 15.18 million, that is, about 14.7 million, which is 1.68 million more than the current permanent resident population, and the family size will be increased by about 730,000 households, which will increase the housing demand of about 730,000 units.

3.2 Forecast of Per Capita Housing Floor Area

Multivariate regression model and Logistic model are used to calculate the per capita housing floor area. In the multiple regression model, per capita housing floor area is expressed as a function of residents' income, urban population and housing price. The per capita housing floor area is expressed as:

$$A = C + C_1 \times I + C_2 \times I^2 + C_3 \times M + C_4 \times M^2 + C_5 \times LogP + U \qquad (6)$$

Where, A is the urban per capita housing floor area, I is the urban per capita disposable income, M is the urban population, P is the sales price of commercial housing, U is the city and year control variable, C is A constant term, and C_1–C_5 is the regression coefficient.

In order to estimate the per capita housing floor area of Shenzhen in 2025, based on the statistical yearbook of different regions, the population size, per capita housing floor area, per capita disposable income of cities and commercial housing sales price of 31 cities were collected, and the Stata software was used for regression analysis. According to the regression analysis, the per capita housing floor area in Shenzhen in 2025 will be about 29 m^2.

3.3 Forecast of Induced Demand

From 2009 to 2015, the establishment stage of urban renewal policy: Shenzhen carried out systematic policy innovation on urban renewal planning compilation, property rights disposal, land transfer, profit distribution, etc., and established the mode of "government guidance, market operation, planning and overall planning, and benefit sharing", thus the development of stock land entered a new stage.

From 2016 to 2020, the implementation phase of urban renewal standards: the role of government guidance was strengthened, and the proportion requirements of public supporting facilities, affordable housing and innovative industrial housing were gradually increased in the renewal.

The trend of urban renewal policy in Shenzhen is as follows: firstly, reasonable planning and allocation of spatial land resources: the plot ratio and transfer rate should be increased simultaneously; Second, the implementation trend of urban renewal over the years: the implementation rate has steadily increased, and the development and implementation are mainly conducted by private enterprises; Third, urban renewal has become more legal and standardized, with more efficient operational procedures.

Combined with the changes of urban renewal policies, the land area for demolition of stock housing in the city since 2006, and the progress of urban renewal projects in the process of application, it is predicted that the demolition area of the city will be about 9 million square meters during the period of 2021–2025 (Table 2).

Table 2. Housing demolition land area in Shenzhen since 2006

Year	Land area for demolition (10,000 m²)
In 2006–2010	133
In 2011–2015	974.7
In 2016–2020	917.4

3.4 Vacancy Rate Prediction

From the point of international practice, the housing vacancy rate in 5%–10% belong to the reasonable interval, due to the housing market, changing need combined with the status quo of housing stock and construction situation, the tendency of the market and the development enterprise intend to factors, such as comprehensive judging the reasonable urban housing vacancy rate during the planning period, on the basis of measuring planning period of the total demand for housing. It is a good way to estimate the vacancy rate of urban housing through housing electricity consumption data to judge whether the housing is vacant or not. At present, the study estimates that the housing vacancy rate is 3% based on the data of sampling survey and human settlement survey.

Based on the above forecasts, the total new housing demand in 2025 can be obtained by substituting each estimate into Formula (2):

$$D_g = (29 \times (1580 - 1344) + 2 \times 1344 + 900)/(1 - 3\%) = 10754.63 \qquad (7)$$

According to 90 m^2/set, the new demand is about 1.2 million sets.

4 Summary

Through the analysis of the new housing demand model, this paper shows that in the next five years, shenzhen will increase the housing demand of about 1.2 million units, among which the net inflow of population is the main force of the new housing demand in the next five years, and the improvement of demand and the induction of demand are also an important part of the new housing demand.

Based on the housing development policy of Shenzhen, it can be seen that the new housing supply capacity in the future is difficult to effectively meet the growth of the new housing demand. So the following policy implications are obtained, which provide certain evidence support and empirical reference for the policy formulation. First, we still need to pay attention to standardize and develop the housing rental market, and strive to solve the housing problem of new citizens. Second, improving the housing security system, increasing the effective supply of public housing, expanding the coverage of public housing and reducing housing costs are important measures to alleviate the imbalance between supply and demand of housing in Shenzhen.

References

1. Chen, S., Ruan, B., Liu, H., Zhang, W.: Positioning and reflection on the development of housing security in Hubei Province during the 14th five-year plan period. Hubei Soc. Sci. **9**, 35–42 (2020)
2. Shan, L., Zhou, Y., Jing, W., Zhou, Z.: Livable urban planning in shenzhen in the new era: exploration and practice. Urban Plan. Rev. **7**, 110–118 (2020)
3. Chen, Y., Chen, X.: The impact of population aging on China's urban housing demand. Econ. Theory Econ. Manage. **5**, 45–57 (2013)
4. Wu, J., Xu, M.: Calculation and analysis of the scale of urban new housing demand in China. Stat. Res. **9**, 75–88 (2021)
5. Deng, H., Huang, G., Xu, S.: Impact of population structure change on housing demand: an empirical analysis based on provincial panel data from 2002 to 2016. J. Central China Normal Univ. Hum. Soc. Sci. **3**, 51–59 (2019)
6. He, R.: Research on the demand of rent and purchase groups in Beijing based on eigenprice model. China's Collective Econ. **29**, 11–14 (2020)
7. Zhu, L., Li, X., Dong, J.: Research on the influencing factors of housing demand under the background of population aging – based on the analysis of population structure, housing price and other factors. Price Theory Pract. **6**, 95–98 (2019)
8. Yang, Z., Li, Q., Hong, J.: Research on the correlation of housing demand and policy impact based on urban agglomeration. East China Econ. Manage. **11**, 100–106 (2019)

Research on the Shenzhen Mode of China's State-Owned Enterprise Reform Serving Industrial Development in the New Era

Yuanyuan Li(✉)

Shenzhen University, Shenzhen 518060, China

Abstract. Based on the neoclassical micro theory, this paper expounds those state-owned enterprises, as the micro main body of the market, have the function of guiding private and small-micro enterprises to fully participate in competition and promote each other's development, analyzes that this function cannot be replaced by other micro main bodies of the market, and emphasizes the role of state-owned enterprises in the development of national economy through the neoclassical micro theory. By collecting and understanding the spirit of the three-year action plan for the reform of state-owned enterprises and the "1 + N" policy system for the reform of state-owned enterprises issued by the SASAC of the State Council in 2020, and in combination with the increasingly cautious supervision of the use of financial funds by the central and local governments, the local governments have accordingly adjusted and changed the investment and financing mode. This transformation is appropriately closely related to the ongoing reform of state-owned enterprises and regional industrial upgrading. It is a challenge and opportunity that local governments must face at this stage to comprehensively promote the vigorous development of regional industries through the reform of state-owned enterprises. As a leading demonstration area of socialism with Chinese characteristics, Shenzhen has formulated an effective reform action plan in the reform of state-owned enterprises and took the lead in implementing the Fund Group Strategy. By adopting the equity investment model, improve the use efficiency of financial funds, comprehensively lead the industrial transformation and upgrading development, to promote the common development of private enterprises. This paper also summarizes the results in the process of practice, in order to provide practical reference and model for the reform of state-owned enterprises, the development of service industry and the development of private economy in the new era.

Keywords: Reform of state-owned enterprises · Full competition · Government investment and financing mode · Equity investment · Fund Group Strategy

1 Introduction

At this stage, China is facing a complex and changeable international environment and multiple challenges. It is necessary to speed up the construction of a new development

Y. Li—Ph.D., worked in China Center for Special Economic Zone Research (CCSEZR), Post-doctoral Workstation for First-level Theoretical Economics, Shenzhen University.

© Springer Nature Switzerland AG 2022
M. A. Serhani and L.-J. Zhang (Eds.): SERVICES 2021, LNCS 12996, pp. 83–97, 2022.
https://doi.org/10.1007/978-3-030-96585-3_7

pattern with the domestic big cycle as the main body and the domestic and international double cycles promoting each other[1], which emphasizes that recycling is not to give up international division of labor and cooperation, but to build China's new international cooperation and competitive advantage based on domestic recycling and relying on the super large-scale market advantage of domestic recycling to attract the resource elements of global commodities.

The sudden outbreak of novel corona-virus pneumonia has seriously affected the global order. China has successfully fought against the epidemic and extended support to all countries in the world to effectively curb the spread of the epidemic in the global area. State-owned enterprises play a key and decisive role in the extraordinary period. At present, in order to meet the needs of current economic and social development and ensure the ability to deal with various unexpected factors and shocks, state-owned enterprises are carrying out self-transformation to meet the development requirements of the new era.

2   Based on the Neoclassical Micro Theory, Analyzing the Main Functions of State-Owned Enterprises in the Market in the New Era

On October 9, 2018, Liu He, leader of the state owned enterprise reform leading group of the State Council, pointed out at the National Symposium on state-owned enterprise reform that "we should understand the central position of deepening state-owned enterprise reform in the new era from a strategic perspective, and focus on promoting six major tasks with the idea of 'breaking one finger rather than hurting ten fingers'." Its strategic significance lies in accurately studying and judging the new changes in the domestic and international environment for the reform and development of state-owned enterprises, understanding the central position of deepening the reform of state-owned enterprises in the new era from a strategic height, and fully understanding the extreme importance of enhancing the vitality of micro market subjects. The meeting[2] stressed that "we should improve the efficiency of state-owned capital allocation through the development of mixed ownership economy. At the same time, we should vigorously support and drive the development of non-public economy, so as to realize the mutual promotion and common development of all kinds of ownership capital."

China's 14th Five Year Plan clearly pointed out that "We will unswervingly consolidate and develop the public sector of the economy and unswervingly encourage, support and guide the development of the non-public sector of the economy. We will accelerate the optimization and structural adjustment of the layout of the state-owned economy and give full play to the strategic supporting role of the state-owned economy. We will accelerate the improvement of the modern enterprise system with Chinese characteristics and deepen the reform of mixed ownership of state-owned enterprises." Since the founding of new China, China's state-owned enterprises have experienced several reforms and development. Some studies believe that the reform of state-owned enterprises at this

---

[1] Liu (2020).

[2] See for details as follows: http://www.gov.cn/xinwen/2018-10/09/content_5328968.htm.

stage is "the country advances and the people retreat", but this statement is one-sided. The reform of state-owned enterprises and the development of private enterprises are a unified whole.

China has gradually opened the market through reform and opening, and formulated various systems to develop the market economy. From the perspective of perfect competitive market, the development of China's market economy has brought two positive effects; On the one hand, develop a competitive market, stimulate market vitality through competition, stimulate each participant to find an appropriate role in the market, promote a more rational distribution of factors and improve the efficiency of resource allocation; On the other hand, a competitive market can ensure the completeness of information, help to judge the real situation of information through commodity prices, and adjust incentive policies accordingly through information feedback. Because there is no transaction cost in the perfect competitive market, the equilibrium price in the market clearing state can reflect all the information of all parties to the transaction. Whether western developed countries regulate enterprises or China reforms state-owned enterprises, the main purpose is to establish a good market environment and promote the full competition of all kinds of enterprises. Therefore, as a micro subject, state-owned enterprises play three functions in the market as below:

2.1 State Owned Enterprises Contribute to the Establishment of an Orderly and Healthy Market Environment

In China's economic development, an economic subject needs to play a long-term, effective, and consistent role in order to ensure the stability of the overall operation of the economy. According to the neoclassical "rational expectation" economic man hypothesis, the essence is that all decisions made by economic man are based on the optimal welfare level of individual (or firm). As a firm, in order to grab the maximum profit in the market, it will make various attempts to achieve a monopoly position. Once private enterprises are in an absolute monopoly position, in order to ensure market share and long-term monopoly, they will obtain excess profits through various means and cause social welfare losses. In this way, they do not use the market to develop healthily. If we only rely on the administrative punishment of the government on monopoly enterprises, we cannot fundamentally solve the problem, and even the potential "Rent-seeking" risk. Due to the natural endowment of private enterprises in the economy (see Table 1), it is impossible for them to undertake the function of promoting the market to achieve the maximum total social welfare. Therefore, in the market, only state-owned enterprises, as one of the micro subjects, play the role of maintaining and establishing an orderly and healthy market environment and play the function of "stabilizer". In fact, in the industries of countries and economies around the world, there are generally enterprises with the state-owned economy dominated by the government as the main nature (Szanyi 2019). The functions and functions of these enterprises in the economy are no different from those of China's state-owned enterprises. After the global economic crisis in 2008, in developed countries such as France, Germany, and the United Kingdom, as well as other countries such as Singapore, Poland, and Turkey, state-owned enterprises play an irreplaceable role in policy implementation.

Table 1. Main development goals of government, state-owned enterprises, and private enterprises in the market

Market role	Overall objectives	Industrial development objectives	Competing objectives	Social responsibility
Government	The total social welfare level is the best	① Industrial structure equilibrium ② Sustainable development	Perfect competition market	Economic growth
State-owned enterprise (SOE)	① Maintenance ② Appreciation state-owned assets	① Resolve excess capacity ② Transfer, upgrade	Guide, support and drive private firms	Supply public goods
Individual Firm	Profit maximization	① Monopoly ② Oligopsony ③ Price discrimination	Monopolistic competition	Cost minimization

2.2 State Owned Enterprises Guide the Market to Fully Compete and Finally Achieve "National Progress"

In the new era, state owned enterprises are self-innovated and fully participate in market competition, to create a fair competition market environment for private enterprises (Xi 2018), and orderly increase market competition level, and help to rectify the distorted price distortions through market mechanism, which is conducive to more complete information and make the parties more accurately judge the real development of the market (Guo 2018). At the same time, by constantly improving the scope and intensity of regulation, the government promotes private enterprises to more standardize their own behavior, and consciously win market share by developing and improving their core competitiveness (See Fig. 1).

As can be seen from Fig. 1, state-owned enterprises play a leading role in orderly guiding all kinds of private enterprises to participate in competition and develop into a fully competitive market. The mixed ownership reform is to improve the competitiveness of enterprises through the deep integration of the advantages of private enterprises and state-owned enterprises, realize the scientific pricing of state-owned assets through the market law, and improve the efficiency of state-owned asset allocation. From the perspective of long-term development of enterprises and industrial transformation and upgrading, mixed ownership is not only the main reform content of state-owned

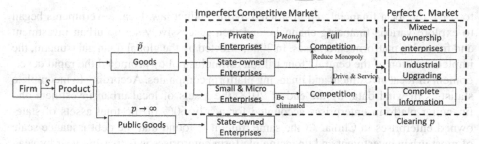

Fig. 1. China's state-owned enterprises guiding private firms to develop into perfect competition market[3]

enterprises, but also an opportunity for all kinds of private enterprises to seek greater development.

2.3 Only State-Owned Enterprises as Public Goods Suppliers in the Market Can Solve the Problem of Market Failure

Public goods are non-exclusive, non-competitive, and external, and their transaction costs are approaching infinity. Competitive manufacturers take profit maximization as the primary production goal, and are unable to continuously produce and provide public goods. According to experience, competitive capital entering the public goods industry will lead to various market failures. Government intervention can only achieve a certain degree of correction, and it is impossible to eliminate this failure. However, the market needs public goods, and the government is not the production department. Therefore, state-owned enterprises can only undertake the task of sustainable production and provision of public goods.

It is worth noting that the mixed reform of state-owned enterprises is based on the "three benefits", which should be conducive to the amplification of the function of state-owned capital, the maintenance and appreciation of value, and the improvement of competitiveness. Once the cart before the horse is turned upside down, it will inevitably lead to the loss of the whole social welfare level and state-owned assets. This is interviewed by FACEBOOK in western developed countries, as the same as the cases of China's Internet platform economy. At this stage, those cases prove that ownership by the whole people and state-owned economy fully show the role of "top beam" and "ballast" in China's economy (Lu 2020), and prove that its role cannot be replaced by private enterprises.

3   With the Transformation of Government Investment and Financing Mode, the Reform of State-Owned Enterprises Presents Main Characteristics

In 1994, since the reform of the "tax sharing system", various expenditures originally borne by the central government were gradually borne by the local government. In order

---

[3] (Remarks: In the Fig. 1, letter S stands for Supply, p stands for price, $\tilde{p}$ stands for equilibrium price, p_{mono} stands for monopoly price).

to solve the local financial gap, restricted by the budget law, local governments began to explore various financing channels, and then successively set up urban investment and financing platform enterprises. In 2008, affected by the global financial tsunami, the implementation of the central "four trillion" stimulus policy promoted the rapid development of local investment and financing platform companies. According to incomplete statistics of wind data, by the end of 2018, the assets of local urban investment and financing platform enterprises accounted for nearly 60% of the total assets of state-owned enterprises in China. At the same time, it is found that the debt issuance scale of most urban investment and financing platform enterprises is increasing year by year. With the potential debt default risk and local debt ratio gradually high. In 2014, GF [2014] No. 43[4] focus on local debt. Thus, the transformation mode of local urban investment and financing platform enterprises also shows that the investment and financing mode of local governments has developed from "bond mode" to "equity mode" (Liu 2021).

Due to different local economic development, in the economically developed eastern region, some local governments took the lead in the transformation of investment and financing platform enterprises to cooperate with the transformation of local government investment and financing mode. The task of western region is usually heavy, with many problems left over by history, and the burden of state-owned enterprises of local governments is heavy. The main problem is to resolve the redundancy and potential risks left over by history.

In October 2018, Chinese Vice Premier Liu he pointed out at the National Symposium (See Footnote 2) on state-owned enterprise reform "Understand the central position of deepening the reform of state-owned enterprises in the new era from a strategic perspective, and focus on promoting the construction of modern state-owned enterprise system with Chinese characteristics, the reform of mixed ownership, the market-oriented operation mechanism, the supply side structural reform, the reform of authorized operation system and the supervision of state-owned assets." Among these six aspects, for the central state-owned enterprises, the key work is the three tasks of enterprise strategic transformation, industrial upgrading of the industry where the central state-owned enterprises are located and realizing the role of the new era of state-owned assets supervision. For the local state-owned enterprises, it is the three tasks of the reform of mixed ownership of equity investment dominated by M & A, the incentive mechanism with modern enterprise characteristics and the exploration of the reform of authorized operation.

In 2020, SASAC of the State Council, through vice premier Liu He, put forward six major tasks to promote the reform of state-owned enterprises in the new era, formulated *the three-year action plan*[5] for the reform of state-owned enterprises according to these six aspects, and on this basis, issued the "1 + N" system for the smooth implementation of the action plan. These policies are the action program for the future development of state-owned enterprises. As the "general construction drawing" and policy pool of state-owned enterprise reform at the present stage, these two series of documents point out

---

[4] See for details as follows: http://www.gov.cn/zhengce/content/2014-10/02/content_9111.htm.

[5] Excerpt from the three-year reform action plan for state owned enterprises on the official website of the state-owned assets supervision and Administration Commission of the State Council.

the direction and objectives of state-owned enterprises for reform. The characteristics of state-owned enterprises at this stage are as follows:

3.1 State-Owned Enterprises are More Responsible for Various National Policies and Arrangements

State-owned enterprises can ensure that national policies are accurate and direct without distortion. Throughout the history of human development, whether the central planner economy or the decentralized economy, the role of the government in the economy is to improve the overall welfare level of the whole society. In the global new coronal pneumonia epidemic, China's state-owned enterprises played a significant role in combating the epidemic, implementing national fiscal policies, and promoting China's economic recovery. State-owned enterprises responded quickly to the instructions given by their superiors, made free directional donations, worked overtime regardless of cost to carry out production and reconstruction, in order to ensure the country's rapid economic recovery under the impact of external emergencies, and once again prove the institutional superiority of China's socialist state-owned economy with the nature of ownership by the whole people (Song 2018)[6]. In 2019, central enterprises became the main investors in major regional development strategies such as the coordinated development of Beijing, Tianjin and Hebei, the integrated development of the Yangtze River Delta, the construction of Guangdong-Hong Kong-Macao Great Bay Area, the construction of Xiong'an New Area and the construction of Hainan Free Trade Zone, involving an investment amount of more than 1.5 trillion yuan, and more than 390 strategic projects signed between central enterprises and local governments. Therefore, only state-owned enterprises can shoulder this important task and implement it consistently when building major projects and implementing major policies across provinces.

3.2 State-Owned Enterprises' Capacity of Supplying and Guaranteeing for Public Goods Has Been Continuously Enhanced

Chinese President Xi Jinping once pointed out at a collective learning conference: "The important industry that promotes the development of social productivity and national security must be the state-owned enterprises owned by the whole people to undertake and invest in development". According to the statistics of the state owned assets supervision and Administration Commission of the State Council, Chinese state-owned enterprises provide public grain, cotton, sugar, salt and other reserve guarantee tasks, ensure the safety of China's rations, build a basic telecommunications network covering all administrative villages and above units in 31 provinces (autonomous regions and cities), provide 95% of the country's crude oil, natural gas and online electricity, and undertake China's

---

[6] Song said: "Enterprise is the most basic carrier of the combination of workers and means of production, and it is the cell that constitutes the economic relationship of ownership by the whole people. Without the support of state-owned enterprises with the nature of ownership by the whole people, there will be no direct combination of workers and means of production, and there will be no real socialist state-owned economy with the nature of ownership by the whole people".

national defense industry All scientific research and production tasks in the fields of national security. Through investment by state-owned enterprises in key areas and public utilities related to people's livelihood, such as pension, medical care, education, and housing, we will fundamentally improve and improve the level of social welfare.

3.3   State-Owned Enterprises Actively Drive the Continuous Optimization of Industrial Structure Layout

At present, as everyone knows that China has encountered the problem of "neck sticking" of key technologies, which needs to overcome difficulties independently. China's state-owned enterprises should actively adapt to and lead the new normal of economic development. Under the "two-wheel drive" of reform, restructuring, innovation and transformation, the development of Finance and strategic emerging industries is an important strategy for the strategic expansion of state-owned assets and state-owned enterprises, and take finance as the main supporting force to serve the real economy and key industries. In terms of driving the development of strategic emerging industries, state-owned enterprises need to play a leading role in key financial support and supporting services.

Whether in western developed countries or China, the government needs to regulate industries. Regulation is given by the government that a series of industries need and control (Stigler 1971), regulate oligarchs in the industry to control the benign and moderate development of the industry, effectively regulate the occurrence of behaviors such as monopoly and government enterprise collusion caused by information asymmetry, moral hazard, and adverse selection. China is experiencing an economic structural slowdown. The supply side structural reform is to adjust the traditional structure formed at the supply side. By adjusting the structure and upgrading the industry, we can promote the rapid improvement of productivity. In 2019,[7] central state-owned enterprises invested 950.5 billion yuan in strategic emerging industries, with a year-on-year increase of 33%, accounting for 20% of the total investment of central state-owned enterprises in that year. The growth rate of investment in new materials industry, energy conservation and environmental protection industry and new energy industry reached 114.9%, 88.1% and 76.8% respectively. Increase cooperation in relevant fields and industries to drive industrial development with investment, Investment leads private enterprises to actively participate. At the same time, central state-owned enterprises accelerated the withdrawal of excess capacity and inefficient and incompetent assets. By the end of 2019, central state-owned enterprises had disposed of 2,041 "zombie" extremely poor enterprises, and the accumulated excess capacity of coal and steel had reached 114 million tons and 16.44 million tons respectively.

3.4   State-Owned Enterprises Playing the Leading Role in Innovation Has Become More Prominent

The economy of developed countries has changed from intensive to innovative growth mode, and people, as the innovation subject, obtain profits through "Knowledge Rent";

---

[7] Data source: Xinhuanet, China news network.

China is in the transition period from factor accumulation to intensive, so we need to pay attention to the efficiency of optimal allocation of factors, and encourage enterprises to carry out technological innovation to promote economic growth into the path of sustainable development (Liu 2018). Technological innovation consumes enterprises' human and material resources, and cannot guarantee the actual benefits in the short term. At present, China is trying independent research and development and independent scientific and technological innovation. As a brave explorer supporting innovation, state-owned enterprises bear the risks of long-term and continuous capital investment and can ensure the needs of technological innovation. In order to overcome difficulties, state-owned enterprises have the courage to take all kinds of risks and make unremitting efforts for China's self-technological breakthrough. From 2016 to 2019[8], state-owned enterprises gradually developed into the main force of the national science and technology innovation strategy. The R & D investment of central enterprises accounted for 26.5% of the national R & D investment, reaching 1.97 trillion yuan. With 733 national R & D platforms and 91 national key laboratories, the leading role of state-owned enterprise innovation in driving social R & D has gradually emerged.

4 Reform Logic of State-Owned Enterprises in Shenzhen

Shenzhen is the first coastal city to rely on "two ends abroad" for foreign trade import and export development in China. In recent years, the Shenzhen Municipal Party committee and government have been fully analyzing the economic development of local governments. In view of the problems arising from local state-owned enterprises, special reforms were carried out for the "iron threshold" accumulated in several reforms. Of course, it is worth noting that Shenzhen, as a special economic zone, is different from the development of state-owned enterprises in economically underdeveloped areas in the West. There are more problems left over from the history of the development of state-owned enterprises in the central and western regions. In contrast, Shenzhen, as a special economic zone of China's reform and opening and a leading demonstration zone of socialism with Chinese characteristics, has embarked on a new road of daring to break through and try.

At the beginning of 2018, the inspection team of Shenzhen Municipal Party committee inspected the Party committee of Shenzhen SASAC[9]. The main problems are: "the overall scale of Shenzhen state-owned assets is small, the comprehensive strength is not strong, and there is an obvious gap with the new requirements of strengthening and optimizing state-owned enterprises." the inspection team wrote in the feedback report: "The total asset scale of municipal state-owned enterprises is still relatively small; the added value accounts for only 5% of the city's GDP, which is far lower than that of Shanghai and Beijing, and its contribution to the city's economy is low, which does not match the urban status and economic volume of Shenzhen".

From 2020, the central government and governments at all levels will adopt a more prudent supervision mode for the use of financial funds. With the process of urbanization,

---

[8] Data source: Xinhuanet, China news network.
[9] See: public information of Shenzhen SASAC.

more and more people will be concentrated in the Yangtze River Delta and Pearl River Delta. The development of Guangdong-Hong Kong-Macao Greater Bay Area can dilute many migrants employed people. For Shenzhen, many urban villages and shantytowns are at the time node that needs to be upgraded in the process of urbanization and renewal, and industrial development also depends more on the actual supporting facilities of the region. The Shenzhen municipal government has realized the urgency and necessity of this development, which is an important engine for maintaining the high growth of the special zone. Shenzhen state-owned enterprises play an important role in it.

Therefore, under the development background of the new era, since 2016, based on the "one body and two wings" (OBTW) development strategy, Shenzhen has established funds with large scale, various types and suitable for different stages to promote the optimization of the layout of state-owned capital and guide the common development of Shenzhen's strategic emerging industries and public affairs, so as to gradually explore a sustainable development model to improve the utilization rate of financial funds, Kill many birds with one stone.

Shenzhen took the lead in adopting the Fund Group Strategy, vigorously developed strategic emerging industries through equity investment and financial funds, developed urban infrastructure construction, and supported private enterprises to grow and expand relying on industries. State owned enterprises in Shenzhen are the main force in the implementation of the Fund Group Strategy. Among them, the fully applied government investment funds include: guidance funds, industrial funds, venture capital funds and other investment forms, and support the investment system covering the whole life cycle of enterprise development. According to the data released by Shenzhen SASAC, by the end of 2020, the number of private equity investment funds managed or participated in by municipal state-owned enterprises exceeded 210, with a total scale of about 420 billion yuan. The establishment of fund groups comprehensively drives and services the synchronous development of strategic emerging industries. On the premise of stable basic business development, the existing state-owned fund supervision enterprises strive to build a fund platform and fund introduction, and support more industrial parks, old city reconstruction and other related sectors, to seek greater development.

5 Specific Measures for Shenzhen's State-Owned Enterprise Reform to Participate in the Fund Group Strategy and Lead Urban Industrial Development

In July 2019, the State Council of China approved the implementation plan[10] for the comprehensive reform of regional state-owned assets and state-owned enterprises in Shenzhen. The plan puts forward: "by 2022, the total assets of state-owned enterprises in Shenzhen will strive to reach 4.5 trillion yuan and build one to two of the world's top 500 enterprises." since the launch of the comprehensive reform experiment, Shenzhen SASAC has continued to promote the development ideas of resource capitalization,

---

[10] See: Shenzhen SASAC website "Implementation plan for comprehensive reform of regional state-owned assets and state-owned enterprises in Shenzhen", http://gzw.sz.gov.cn/zwgk/zcfgjz cjd/zcfg/content/post_4562753.html.

asset capitalization and capital securitization, and accelerated the implementation of the strategies of "Listed Companies +" and "+ Listed Companies". At present, the asset securitization rate of state-owned enterprises in Shenzhen has reached 57.8%, and the overall completion of quantitative indicators is close to 80%[11].

In 2020, the global epidemic broke out, the logistics chain was at a standstill, and the factors could not operate at the same high speed, which restrained the export trade. Shenzhen government makes many useful arrangements just in time, that formulate the policy of "serving the overall situation, serving the city, serving the industry and serving the people's livelihood" for financial assistance to urban development. Under the impact of multiple pressures and the condition of ensuring existing advantages. During the 14th Five Year Plan period, Shenzhen will fully focus on turning weaknesses into development advantages.

The municipal state-owned funds group has continuously improved the whole process innovation ecological chain of "basic research + technological breakthrough + achievement industrialization + science and technology finance", to inject strong power into Shenzhen to accelerate the high-quality innovation and development of high-tech industries and better play the role of demonstration and driving. According to the latest data released by Shenzhen SASAC, it pays more attention to the functional positioning of financial service entities. The principle of "three investment and seven service". By the end of 2020, the total assets, net assets, operating income, total profit, net profit, and taxes paid by municipal state-owned assets had reached "six new highs", of which the total profit was 135.1 billion yuan, an increase of 5.3%, and the net profit was 102.6 billion yuan, an increase of 7.8%. Compared with before the comprehensive reform test, the total assets increased by 31.8%, the operating revenue increased by 59.1% and the total profit increased by 25.8%. At this stage, many state-owned enterprises in Shenzhen fully serve the mixed reform of state-owned enterprises, industrial upgrading, layout of strategic emerging industries and other key work (see Table 2).

As shown in Table 2, Shenzhen state-owned enterprises are implementing the Fund Group Strategy. The specific conditions and industries of each enterprise are as follows:

Shenzhen Investment Holdings Co., Ltd. (SZIHC)[12] is mainly concentrated in three sectors: Science and technology finance, science and technology industrial park and science and technology industry. It controls a number of leading enterprises in subdivided industries such as Guosen Securities[13], Shenzhen High Tech Investment Group (SZHTi)[14], Shenzhen Credit Guarantee Group (SZCGC)[15], etc., holds 10 listed companies, and mainly participates in China Resources International Trust[16], Kunpeng Capital[17], China state-owned capital venture capital fund, Guotai Junan Securities (GTJA)[18],

---

[11] Data source: Shenzhen SASAC website.

[12] See: http://www.sihc.com.cn/en/index.php/enhome.

[13] See: https://www.guosen.com.cn/gs/openaccount/?qrcode=1563263945301&aid=156326394 5301&bd_vid=8425685925147660134.

[14] See: https://www.szhti.com.cn/#/big/company?id=0c1eac83a44942ebbbec5ec6ca356b55& name=%E5%85%AC%E5%8F%B8%E7%AE%80%E4%BB%8B.

[15] See: https://www.szcgc.com/.

[16] See: https://www.crctrust.com/gsjj/index.html.

[17] See: http://www.kpcapital.cn/.

[18] See: https://www.gtja.com/content/gtja_en/home.html.

Table 2. Shenzhen state-owned enterprises' participation in the strategic service industry of Fund Group

No	Enterprise name	Fund's name	Service industry category
1	SZIHC	① Angel Funds; ② VC/PE Funds; ③ M & A Funds	① Science and Technology Park; ② Finance; ③ Public Utilities and Services
2	Shenzhen Capital Group Company	① Venture Capital Fund; ② Guiding Fund; ③ Industry Funds	① Science and Technology; ② Urban Infrastructure; ③ Agricultural Products
3	Shenzhen Angel FOF	① FOF; ② Venture Capital Fund; ② Sub-funds;	① New Generation Information Technology; ② Biomedical Industry; ③ High-end Equipment Manufacturing
4	SZHTi& SZCGC	① Talent Fund; ② Venture Capital Fund;	① Talents Plan; ② Service Small and Medium-sized Enterprises
5	Shenzhen Capital Holdings Corporation	Equity Investment Funds;	Private Economic Development
6	Shenzhen Major Industry Investment Group	Major Industry Development Funds	① Advanced Manufacturing; ② Emerging Industries; ③ LCD Panel Industry
7	SZIHC	Infrastructure Investment Funds	① Construction of Urban Infrastructure; ② Public Services
8	Shum Yip Group	① Hou An Innovation Fund; ② Early and Growth Investment	① Chip Design; ② Patents; ③ Intellectual Property Rights
9	Shenzhen Talents Housing Group	REITs;	Talent Rental Housing
10	Shenzhen Bus Group	Big Transport Industry Fund	① New Energy Vehicles; ② Ground Transportation Industry ③ Urban Transportation

China Ping An Insurance[19], Shenzhen Energy[20], China Merchants Bank[21], etc. it has established Angel Funds, VC / PE Funds, M & A Funds and other fund groups covering the whole life cycle of enterprise development. As a technology and financial holding platform, investment holding companies actively carry out systematic fund operation, build a technology industry investment fund group, promote the operation of the linkage business model, and set up 42 funds throughout the system, The total capital scale is 81.1 billion yuan.

Shenzhen Capital Group Company[22], it is the first venture capital enterprise engaged in equity investment. It was established early and has rich project operation experience, ranking among the best in the national venture capital industry. Shenzhen Guiding Fund managed by the group guides social capital to invest in urban infrastructure construction, livelihood development and other fields. Shenzhen Gas[23] and Shenzhen Agricultural Products Group (SZAP)[24] respectively use or set up relevant Industrial Funds to promote the development of urban gas and agricultural products circulation industry and serve urban infrastructure and people's livelihood.

Shenzhen Angel FOF[25], is a venture capital enterprise engaged in high-risk investment. It is a major policy measure to benchmark international first-class enterprises, supplement the short board of venture capital, and help enterprises in seed stage and start-up stage; With a registered capital of 10 billion yuan, it is the largest Angel FOF in China. The first phase of 5-billion-yuan municipal Angel FOF was established to improve the innovation ecosystem and achieve practical results. From March 2018 to January this year, the total scale of Angel FOF of Sub-funds was RMB 6.4 billion. Angel master fund promised to invest RMB 2.56 billion and completed the investment of RMB 563.5 million to 11 companies. The eight sub-funds invested considered and approved 21 projects to be invested, with a total investment of about RMB 180 million. The investment proportion of the project is up to 40%, and all excess income is transferred. At present, the total number of investment decision-making sub funds amount has exceeded 50, and the sub funds have invested in more than 270 Angel projects, of which the number of projects in the new generation information technology industry, biomedical industry and high-end equipment manufacturing industry ranks among the top three.

The three talent innovation and entrepreneurship funds managed and operated by Shenzhen Capital Group Company (also called **SZVCG**), **SZHTi** and **SZCGC** (not former one, is Shenzhen Credit Guarantee Group) respectively, with a total scale of 10 billion yuan, are the largest Talent Fund in China, with a cumulative investment of more than 2 billion yuan and more than 110 investment talent projects, covering a number of national thousand talents plan, peacock plan, Shenzhen overseas high-level talents Shenzhen's high-level talents and other professionals in various fields effectively support

---

[19] See: https://www.pingan.com/official/home-a?channel_id=1.

[20] See: http://www.sec.com.cn/.

[21] See: http://www.cmbchina.com/.

[22] See: http://www.szvc.com.cn/english/Home/index.shtml.

[23] See: https://www.szgas.com.cn/.

[24] See: https://www.szap.com/aboutUs.

[25] See: http://www.tsfof.com/.

the innovation, entrepreneurship and development of talents in Shenzhen. Meanwhile, in terms of serving small and medium-sized enterprises, among the projects invested by the three above companies, CHIPSCREEN[26], JPT[27], Lifotronic[28], FSQuality[29] and other enterprises have been listed. Shenzhen Venture Capital, high tech investment and small and medium-sized enterprises have helped enterprises raise 767.7 billion yuan, boosting 658 enterprises to list on the new third board and 368 enterprises to list on the new third board.

Shenzhen Capital Holdings Corporation[30], Kunpeng Capital and Guosen Securities invested 5 billion yuan to attract various types of capital to set up Equity Investment Funds for listed companies with a total scale of 20 billion yuan in accordance with the policies of the municipal Party committee and the municipal government to service private economic development, which effectively alleviated the liquidity risk of some private listed companies in Shenzhen through equity support.

Shenzhen Major Industry Investment Group[31] has established a Major Industry Development Fund to support major industrial projects in Shenzhen through market-oriented operation, and promote the development of major projects such as advanced manufacturing and emerging industries with large total investment and strong driving effect. It has invested 8 billion yuan in SZCSOT[32] G11 project to promote the upgrading of LCD panel industry.

Shenzhen Construction and Development Group[33], SZIHC and Shenzhen Capital Group Company jointly established a professional fund management company to manage the first 100 billion Shenzhen Infrastructure Investment Fund and leverage social capital to participate in the construction of urban infrastructure and public services.

Shum Yip Group[34] participated in the establishment of Hou An Innovation Fund and joined hands with ARM, the world's leading provider of chip design, patents, and intellectual property rights, to focus on early and growth investment in cutting-edge technology fields.

Shenzhen Talents Housing Group[35] and Shenzhen Capital Group Company jointly initiated the establishment of the country's first and largest talent rental housing REITs (Real Estate Trust and Investment Fund) project, with a total amount of 20 billion yuan. The first phase of 3.1 billion yuan was officially listed and traded on the Shenzhen Stock Exchange, organically combining fund innovation and housing projects, to explore the "Shenzhen model" and development of public housing investment and financing for the whole country "Shenzhen example".

---

26 See: https://www.chipscreen.com/.
27 See: http://www.jptoe.com/.
28 See: http://www.lifotronic.com/.
29 See: http://www.fujipcb.cn/.
30 See: https://www.szcapital.com.cn/#/home/index.
31 See: http://www.sznews.com/content/mb/2021-03/02/content_24011550.htm.
32 See: http://www.szcsot.com/.
33 See: https://www.sztqjf.com/about.aspx?tags=3.
34 See: https://www.shenyejituan.com/.
35 See: https://www.szrcaj.com/.

Shenzhen Bus Group[36] has set up the Big Transport Industry Fund to actively grasp the new market of transportation and new opportunities for new energy vehicles, intelligent transportation, Internet plus traffic and industrial upgrading, and invest in the merger and consolidation of Shenzhen's ground transportation industry, as well as new technologies related to transportation, new energy related electric vehicles, charging, battery, energy storage, travel information and other upstream and downstream related issues. The new model of innovative enterprise equity investment will promote the integration and revitalization of the city's public transport industry resources and serve the development of urban transportation.

Therefore, Shenzhen has successfully leveraged social capital by adopting the Fund Group Strategy, amplified the function of state-owned capital, and actively supported the development of private enterprises and other enterprises of various ownership. It also efficiently through the practice combines capital with resources required for industrial development, the operation of the linkage business mode of "Introduction of Scientific and Technological Innovation Resources + Science and Technology Industrial Parks + Science and Technology Finance + Listed Companies + Industrial Clusters" will be gradually formed to meet the needs of industrial development in the new stage.

References

Liu, H.: "Accelerating the construction of a new development pattern with domestic circulation as the main body and domestic and international double circulation promoting each other", excerpted from the website of the Central People's Government of the People's Republic of China (2020). http://www.gov.cn/xinwen/2020-11/25/content_5563986.htm

Liu, X.: Structural changes and economic growth in China's 40-year of reform and opening up. J. Yangzhou Univ. (Hum. Soc. Sci. Ed.) **2018**(06), 5–17 (2018)

Liu, X.: Chinese urbanization. J. Hunan Univ. **2021**(5), 34–49 (2021)

Lu, Y.: Give Full Play to the Important Role of State-Owned Enterprises in Economic Growth. Economic Reference Daily, 17 May 2021, (05), Edition 007 (2020)

Guo, F.: Historic Achievements in the Reform of State-owned Enterprises (Commemorating the 40th Anniversary of Reform and Opening up). People's Daily Theory Edition, 19 November 2018, (11), Edition 009 (2018)

Song, F.: On the consistency of 'state-owned enterprises become stronger, better and bigger 'and 'state-owned capital become stronger, better and bigger'. Polit. Econ. Rev. **2018**(03), 3–15 (2018)

Xi, J.: Vigorously support the development and growth of private enterprises. In: Xi Jinping'sThird Volume of Governing the Country, vol. 2018, no. 11, pp. 263–268. Foreign Language Press (2018)

Szanyi, M.: Seeking the Best Master: State Ownership in the Varieties of Capitalism. Central European University Press (2019). Chapter 1, 2, 3, 5, 8, 9,10

Stigler, G.: The theory of economic regulation. Bell J. Econ. **2**(1), 3–21 (1971)

---

[36] See: http://www.szbus.com.cn/.

Ensuring People's Health is a Strategic Development Priority

Kunjing Zhang(✉)

Shenzhen Institute of Information Technology, Guangdong 518172, People's Republic of China
2013100916@sziit.edu.cn

Abstract. Xi Jinping's important exposition on people's health has always given priority to the protection of people's health in the strategic position of development, which fully reflects the ruling concept of "people centered" of the Communist Party of China. Starting from the inevitable requirement of building a moderately prosperous society in all respects, Xi Jinping pointed out that "there will be no moderately prosperous society in all respects without the health of the whole people". Starting from the significance of ensuring people's health, the author points out that "people's health is an important symbol of a prosperous nation and a prosperous country". From the perspective of ensuring the implementation of the people's health policy, it is pointed out that "strengthening the institutional guarantee of improving the people's health level"; Based on the basic principles of dealing with major public health hazards, it is pointed out that "people's life safety and health should always be put in the first place".

Keywords: People's health · Healthy China · Xi Jinping Thought on Socialism with Chinese Characteristics for a New Era

1 Introduction

The important statement on people's health in Xi Jinping Thought on Socialism with Chinese Characteristics for a New Era always gives top priority to ensuring people's health, which fully demonstrates the CPC's people-centered governance philosophy. At the Fourth Session of the 13th National Committee of the Chinese People's Political Consultative Conference (CPPCC) in 2021, General Secretary Xi Jinping once again stressed that: "To protect people's health in the priority development of strategic position, adhere to the basic medical and health cause of public welfare, focusing on the major diseases and the main problems affecting people's health, to speed up the implementation of health action in China, woven tightly to national public health nets, pro-mote the development of public hospital quality, provides the Omni-directional whole cycle health services for the people." [1] to Xi Jinping, on the basis of the important statements about people's health, security analysis of a new era of the people's health on the significance and principle of priority to the development of strategic position, has some reference value for further deepening the research on people's health, for our learning, correct understanding and the research on people's health, in turn, scientific. Thinking and observing real problem, it is of practical significance to consciously resist the "western theory of human rights" and to strengthen the "four assertiveness".

© Springer Nature Switzerland AG 2022
M. A. Serhani and L.-J. Zhang (Eds.): SERVICES 2021, LNCS 12996, pp. 98–105, 2022.
https://doi.org/10.1007/978-3-030-96585-3_8

2 Results and Discussion

2.1 Without Universal Health, There Will be No Well-Off Society in All Respects

Good health is the foundation for all good lives, Xi said, stressing that "without health for all, there is no well-off society in all respects." [2] this discussion fully reflects the CPC's accurate grasp of the main contradiction of the Chinese people at the present stage. At the present stage, the main contradiction of the Chinese people has changed from "whether" in the past to "good or not" at present. Along with the Chinese people's yearning for a better life, health has been paid more and more attention by people. It also guided China to firmly establish the concept of national health in the practice of governance in the new era.

(1) **We have firmly established the concept of national health in china**

Under the guidance of Xi Jinping's concept of people's health, more and more people have joined in the national fitness campaign, which is many times better than building dozens of more hospitals nationwide because of its low input and high output. Now a days, "I exercise, I am healthy", the concept of national fitness that sports can make life better has been deeply rooted in the hearts of the people. Everywhere in the motherland, we can see the national fitness campaign such as square dancing in full swing, and the national health level is being improved as a whole. Only when the Chinese people are in good health will they have ample energy to build the country, complete the building of a moderately prosperous society in all respects in a fast and sound manner, and add strong impetus to the realization of the Chinese Dream of national rejuvenation.

(2) **We fully promoted the effective implementation of the healthy china strategy**

The building of a healthy China carries the common ideal of the CPC and the people, and is the expectation and call of the people for the health of the whole people in the new era. The Healthy China strategy will ensure that all Chinese people enjoy equal access to basic medical and health services, fully protect the basic health needs of the Chinese people, promote the development of China's health service industry, and meet the diverse health needs of the people. The Communist Party of China and the government will continue to comprehensively promote the effective implementation of the Healthy China Strategy for the realization of the health of the whole people, because the health of the people is the basic guarantee for the pursuit of a happy and better life, and the overall well-off society inevitably requires the health of the whole people.

2.2 People's Health is an Important Sign of a Prosperous Nation and a Strong and Prosperous Country

"People's health is an important symbol of a prosperous nation and a strong and prosperous country," Xi said, referring to the importance of ensuring people's health. [3] this discussion fully reflects the development concept of "people first" of the Communist Party of China. China's economic development has always given priority to people's health and pointed out the importance of ensuring people's health. In the fight against COVID-19, the Communist Party of China led the Chinese people to save every life at all costs, successfully blocked the spread of the epidemic, and became the first in the world to achieve positive economic growth from negative. China has enabled the world to witness not only the rapid economic development of China, but also the healthy development of the Chinese people. At the same time, it has also witnessed the vigorous development of industries in China that are suited to health.

(1) **It fully demonstrates the CPC's respect for the Chinese people's right to health**

"The right to health is an inclusive and fundamental human right, a fundamental guarantee of a dignified human life and the right of everyone to the enjoyment of the highest standard of health that is equitable and accessible." [4] in the past for a long time, the Chinese people's right to health is can not get effective guarantee, but as China's comprehensive national strength increasing, the healthy development of the Chinese people's index gradually moving towards the world, China gradually explore formed fits a pattern of China's national conditions and health care, and achieved remarkable results. Take the average life expectancy, the best indicator of people's health, as an example. The average life expectancy in China has doubled since 70 years ago. Statistics show that when the People's Republic of China was founded, the average life expectancy was only 35 years old. By 2018, the average life expectancy in China had reached 77 years old, four years higher than the world average. The change in the average life expectancy of the Chinese people fully demonstrates that the healthy development strategy of the Chinese people has brought tangible health benefits to the people. It also shows that the country is strong and prosperous and the nation is prosperous.

(2) **It has vigorously promoted the vigorous development of china's health industry**

In xi "health in all policies, the people sharing" [5] thought under the guidance of local governments in accordance with requirements of the reform of "pipes", further optimize the policy environment, eliminate the institutional obstacles, support and encourage the many social enterprises provide health services for the people, strongly promote the development of China's health industry. The vigorous development of the health industry has provided all-round support and protection for people's health. For example, people can sign a contract with a family doctor in their own community, enjoy door-to-door professional medical services without going out of their homes, and enjoy online consultation services and other convenient medical services. With the rapid development of China's health industry in the future, the Chinese people will be healthier, China will be more prosperous, and the Chinese nation will be more prosperous.

2.3 Strengthening Institutional Guarantee for Improving People's Health

In order to better meet the people's growing needs for a better life, Xi said, "We will strengthen the institutional guarantee to improve people's health," starting from the implementation of the health policy [6]. The improvement of people's health in the form of system fully embodies the CPC's firm confidence and determination to safeguard people's health, and also fully embodies the common aspiration of the people of all ethnic groups in China.

(1) **It fully demonstrates the firm determination of the party and the government to safeguard people's health**

Only the continuous improvement of the people's health level can be conducive to the harmonious and stable development of the society. Since the 18th CPC National Congress, China has formulated and implemented a series of government documents, including the 13th Five-Year Plan on Health and Health, the National Fitness Plan (2016–2020), and the Outline of the Healthy China 2030 Plan, in order to better strengthen and improve the institutional guarantee of people's health, which has brought people's health to a new level. As a result, the main health indicators of Chinese residents are generally better than the average level of middle and upper income countries. By the end of December 2020, the Ministry of Finance, the State Administration of Medical Insurance, and the State Administration of Taxation had helped "increase the per capita financial subsidy for basic medical insurance for residents from 240 yuan in 2013 to 550 yuan in 2020, raising the serious disease insurance from zero, benefiting tens of millions of people." [7] In 2021, provincial pooling of national medical insurance and cross-provincial direct settlement of outpatient expenses will be basically realized. At the same time, the government has also noticed that with the economy, environment and resources of our people's health, food, jobs, education and other fields is closely related to the industrialization, urbanization and ecological environment and lifestyle changes, such as bring to people's health level of new challenges, the party and the government is from the governance level as a whole, We will work hard to solve these long-term and major problems concerning people's health at the institutional level.

(2) **We have effectively ensured that all the people across the country enjoy systematic and continuous health services**

The health of the people has a bearing on the long-term stability of the country and the happy life of thousands of families. It is the common aspiration of all ethnic groups in China to have high-quality and inexpensive health services. From the party's 19th National Congress report requirements "to improve the national health policy, to provide the people with a full range of full cycle health services." [3] from the decision of the Fourth Central Committee of the 19th CPC to emphasize "enabling the broad masses to enjoy fair, accessible and systematic and continuous health services" [6], the CPC is committed to the fundamental purpose of ensuring and improving the people's health level from the institutional level. In order to effectively implement the system and policy of improving the people's health and benefit the people of all ethnic groups, the Party and the government have adopted many effective measures, such as deepening the reform of the medical and health system in an all-round way and safeguarding the public welfare of public health care. The Chinese government is committed to promoting the transition from the "demographic dividend" to the "health dividend", ensuring the health of all ethnic groups in China in an all-round way, and laying a solid institutional foundation for the smooth realization of the great rejuvenation of the Chinese nation and the "two centenary goals".

2.4 Always Put the Safety and Health of the People First

Since the outbreak of novel coronavirus pneumonia in Wuhan and other places in Hubei province in 2020, Xi has attached great importance to the epidemic. Based on the basic principles of dealing with major public health hazards, he stressed that "we always put the safety of people's lives and health first" [8]. This discussion fully clarified the people-centered core of Xi Jinping Thought on Socialism with Chinese Characteristics for a New Era, and clarified the basic principles for dealing with major events that endanger public health.

(1) **First of all, the basic guidelines for the prevention and control of the epidemic have been clarified**

Under the overall direction of the CPC Central Committee with Xi Jinping at its core, Party committees and governments at all levels have resolutely implemented the guiding principles and arrangements of the CPC Central Committee, and united as one. In the face of the sudden outbreak, the CPC Central Committee immediately directed the whole country to make every effort to fight the epidemic and ensure the safety and health of the people at all costs. Second, more than 42,000 medical workers across the country, despite the risk of death, have helped Hubei and Wuhan fight the epidemic together. In the difficult time of fighting the epidemic, General Secretary Xi sent his care and greetings to the medical workers on the front line of fighting the epidemic. After hearing General Secretary Xi's encouraging words, the soldiers in white became more confident and brave, and took practical actions to fight the epidemic to the end, fulfilling their mission. In addition, the broad masses of the people united in fighting the epidemic. "Prevention and control of the epidemic is a serious struggle to protect people's lives and health" [9]. In order to fight the epidemic, people all over the country and overseas Chinese have donated money and materials, and made efforts to coordinate resources from all over the world to purchase urgently needed protective clothing, masks, protective glasses and other materials for the prevention and control of the epidemic, and donated them to Wuhan, the worst-hit city. Some people at the grassroots level have made contributions in their own way, such as volunteering as shuttle workers and supplies for epidemic prevention and control, volunteers for epidemic investigation, and providing free food and accommodation to medical workers. This epidemic fight, let us once again deeply felt the unity of the Chinese nation, once again deeply felt the importance of the CPC Central Committee to the safety of people's lives and health.

(2) **This fully embodies the original aspiration and mission of Chinese Communists**

Since the outbreak of the epidemic, the Standing Committee of the Political Bureau of the CPC Central Committee has held three major meetings in just 19 days to study epidemic prevention and control, with the primary goal of saving as many lives as possible at all costs. On the one hand, the CPC Central Committee coordinated all the forces and gave full support to Wuhan and Hubei to save the lives and health of the people as much as possible. On the other hand, all those who are derelict in the prevention and control of the epidemic will be held seriously accountable, demonstrating the advanced nature and purity of the Party. In the fight against the epidemic, Party members and the general public were heart to heart and worked together to fight the epidemic and finally won the battle against the epidemic. In the fight against the epidemic, a number of outstanding Party members and touching stories have emerged, which once again deeply impressed the people with the original aspiration and mission of the CPC.

(3) It forcefully countered the fallacy of "western theory of human rights"

Western human rights is a general term for the basic views of the western bourgeoisie on human rights. In contemporary times, Western capitalist countries often use human rights issues to interfere in the internal affairs of other countries, promote their values, ideologies, political standards and development models, and damage the sovereignty and dignity of developing countries [9]. In recent years, some Western countries, led by the United States, often use human rights as an issue to violently interfere in China's internal affairs. Through this fight will be coronavirus outbreaks, let us recognize the, Chinese is the most human rights, the Chinese government is the respect for human rights, because in the life and health of people are threatened right and the right to health is the basic human rights, ordinary people's right to exist and the right to health and many western countries, are not guaranteed in the outbreak. As of April 21, 2020, the cure rate in the United States is 9%, while "more than 77,000 COVID-19 cured patients have been discharged from the hospital in our country, with a cure rate of approximately 94% [10]. These data fully demonstrate the Chinese government's respect for the people's rights to life and health. In western countries, the so-called human rights have always been a double standard, which is mainly to protect the bourgeois rule and interests. In the United States, human rights are only for those who are rich, such as the novel coronavirus test. The first to be tested and treated are those who are rich, and the poor, because they cannot afford the test, have to self-quarantine and wait to die even if they are infected with the virus. In the UK, the government had taken the absurd approach of infecting 60% of the population in the early stages of the pandemic in exchange for total immunity. In the West, due to the lack of anti-epidemic materials, it is impossible to take care of all the people when treating the patients. Therefore, selective treatment has been adopted, giving priority to the treatment of young people, while the elderly infected with the virus will be selectively abandoned. For example, we have also seen in the news that Novel Coronavirus infection in American nursing homes has a higher disease death rate. In contrast, in China, all citizens are well treated, from infants to 108-year-old years old, and all are free of charge. So after the fight against the epidemic, many Chinese people are feeling that "I have no regrets in this life to enter China, and I will still be a Chinese in the afterlife.

(4) Strengthening the "four confidences" of the people

In the face of similar public health crises, the Chinese government's rapid response and treatment efficiency are better than those of other countries, fully demonstrating the unique advantages of the socialist system with Chinese characteristics. The COVID-19 epidemic occurred just before the Chinese New Year. In order to better protect people's lives and health, the Chinese government called on 1.4 billion Chinese people to quickly enter the epidemic prevention state, and all provinces and cities launched the level 1 response of major public health incidents. It only took 10 days to build the Huoshenshan Hospital with a total construction area of 33,900 square meters and a capacity of 1,000 beds in Wuhan, and it only took 12 days to build the Rayshenshan Hospital with a total construction area of 79,700 square meters and a capacity of 1,600 beds, all of which were used for the treatment of critical patients. These effective measures have significantly increased the admission rate and cure rate of patients infected by the epidemic, and significantly reduced the death rate and infection rate. From the cities to the countryside, Chinese people have heeded the call of the CPC Central Committee to stay indoors in isolation to prevent the spread of infection and the epidemic. Such speed and efficiency are not possible in other countries. For example, as the most developed capitalist country in the world, the performance of the United States in the fight against the epidemic has greatly surprised people around the world. For some time, the United States has led the world in the number of cumulative cases and deaths of COVID-19, and as of September 8, 2020, the United States has led the world in the number of cumulative cases and deaths of COVID-19. At the same time, there were months of anti-racial riots in the United States, and the Trump administration responded by using force to suppress them, resulting in more American casualties and virus infections. It is conceivable that if the US government had put the safety and health of its people first, it would not have caused such a huge loss. Therefore, the COVID-19 epidemic prevention and control campaign has strengthened the confidence of the Chinese people in the theory, path, system and culture of socialism with Chinese characteristics compared with those of other countries.

3 Conclusion

To sum up, Xi Jinping's important remarks on people's health adhere to the original aspiration of the Communist Party of "seeking the happiness of the Chinese people", and have always placed people's health at the strategic position of priority in development. To Xi Jinping, on the basis of the important statements about people's health, according to the "overall research to deep research to fine," the principle of deepening new era Xi Jinping, the thought of socialism with Chinese characteristics to the people as the center to study the thought of, for our learning, correct understanding and the research on people's health theory and scientific thinking and observing reality problem, resist consciously "western human rights", It is of great theoretical and practical significance to strengthen the "four self-confidence".

Acknowledgements. This research was supported by Guangdong Provincial Department of Education "Eight Unified" University Ideological and Political Course Construction Demonstration Project.

References

1. Xi Jinping visited members of the medical and health education circles who attended the CPPCC meeting. Guangming Daily 2021–3–7(01)
2. National Development and Reform Commission: The 13th Five-Year Plan for National Economic and Social Development of the People's Republic of China. People's Publishing House, Beijing p. 543 (2016)
3. Xi, J.: Decisive Victory in Building a Moderately Prosperous Society in an All-round Way and Achieving the Great Victory of Socialism with Chinese Characteristics in the New Era – Report at the 19th National Congress of the Communist Party of China, Beijing, People's Publishing House vol. 54, p. 38 (2017)
4. The State Council Information Office of the People's Republic of China. The Development of Health Cause and Human Rights Progress in China. Beijing, People's Publishing House, vol. 7 (2017)
5. The 19th National Congress of the Communist Party of China Report Guidance Reader. Beijing, People's Publishing House, vol. 447, no. 5 (2017)
6. Decision of the Central Committee of the Communist Party of China on Several Major Issues of Adhering to and Improving the Socialist System with Chinese Characteristics and Promoting the Modernization of the National Governance System and Governance Capability, Guidance Reader. Beijing, People's Publishing House, vol. 39, p. 302 (2019)
7. Safeguards people's health and builds a healthy China – General Secretary Xi Jinping's important speech at the joint meeting of medical and health education committee members of the Chinese People's Political Consultative Conference (CPPCC) aroused warm response. Economic Daily, 8–3–2021
8. Xi Jinping on People as the Center (2020) [EB/OL] (03–02–2020) [06–08–2020]. https://www.xuexi.cn/lgpage/detail/index.html?id=12790482541700099510
9. Sun Guohua, Ed. Chinese Law Dictionary Jurisprudence Volume Beijing, China Procuratorial Press, vol. 450 (1997)
10. National Health Commission: Our country COVID 19 cure rate more than 94% more than 77000 patients have been discharged [EB/OL]. (21–04–2020) [06–08–2020]. http://news.youth.cn/gn/202004/t20200421_12297302.htm

From Digital Resource Construction to Public Cultural Services: The Innovation Approaches of Digital Culture Construction of Public Libraries

Yi Li[1(✉)] and Ting Xu[2]

[1] Shenzhen Institute of Information Technology, Shenzhen 518172, Guangdong, People's Republic of China
[2] Guilin Library of Guangxi Zhuang Autonomous Region, Guilin 541100, Guangxi, People's Republic of China

Abstract. In recent years, public libraries have undertaken an important social responsibility in digital resource construction and public cultural services. Based on summarizing the development status of digital culture construction of public libraries, it was found that there were contradictory and inconsistency in the aspects of the relationship between supply and demand of resources construction, supply-demand relationship in resources construction, resources integration, service effect and efficiency, marketing promotion and social participation. Through learning from successful experiences and models of public cultural institutions at home and abroad, this paper has discussed several innovative approaches in digital cultural construction of public libraries. Firstly, based on the needs of readers, to promote the supply of digital resources. Secondly, to strengthen resource integration and to work on the perfection of the construction of 'one-stop platform'. Thirdly, to establish a reader feedback mechanism and to customize personalized services. Fourthly, to carry out cross-border library cooperation of 'Internet +' to attract public participation. Finally, to build the library brand IP and to value much more the creativity of service marketing.

Keywords: Digital resource construction · Public cultural services · Public libraries · Digital culture construction

1 Introduction

In November 2020, 'Opinions on Promoting the High-quality Development of Digital Culture Industry' was issued by Ministry of Culture and Tourism which required to cultivate new formats of digital culture industry. There were two important suggestions for public libraries: first, to make excellent cultural resources 'alive' with the aid of digital technology, it was needed to promote the digitization of excellent cultural resources. Second, what calls for special attention is deepening convergence and development to promote the application of digital cultural products and services in public cultural venues, and to enrich the experience form and content of public cultural space [1]. It means that

M. A. Serhani and L.-J. Zhang (Eds.): SERVICES 2021, LNCS 12996, pp. 106–112, 2022.
https://doi.org/10.1007/978-3-030-96585-3_9

the future development direction of public libraries should be to integrate public digital cultural resources and to innovate public cultural services.

2 Literature Review

2.1 The Construction of Public Digital Resources

In recent years, the provincial public libraries (i.e., the three cultural huimin projects, those were the sharing project of national cultural information resources, the popularizing project of digital library and the construction plan of public electronic reading rooms have built a large number of excellent digital culture resources through public digital culture project. The types of construction of resources include special resources of local characteristics culture, special resources of local revolutionary culture, open classes of libraries, special resources of digital libraries, the digitization of local literature and audio and video resources of local characteristics, etc. For example, Guilin Library of Guangxi Zhuang Autonomous Region has successively developed a series of feature projects, such as 'Bagui Historical and Cultural Famous City', 'Guilin Historical Relics', 'Guilin Landscape Culture' and 'Guilin Ancient Villages'. Fujian Provincial Library has built a series of "Fujian Cultural Memory Media Repository", such as "Fuzhou Qiu Hengming's Chong Yao Jin Works" and "Pearly Celadon Culture of Song Dynasty". Shanxi Provincial Library has produced 15 series of cultural feature films with more than 300 collections, such as Ancient Villages of Shanxi, Ancient Great Wall of Shanxi, and Ancient Architecture of Shanxi. Shanxi Provincial Library has built a series of multimedia resource repositories with themes such as Shanxi Qinqiang Opera and Qin Rhyme, Red Memory of Shanxi-Gansu-Ningxia Boundary Region, Shanxi Emperor's Mausoleum, and Silk Road, as well as cultural feature films such as The Silk Road Folk Song of Ming and Qing Dynasties, Xi'an Drum Music, Tang Mausoleum in Guanzhong, and Qin Chuan Buddha Rhyme. Most of these digital resources reflect the local characteristics and folk culture of the region, and play a very critical role in digging and protecting the folk intangible cultural heritage, inheriting and developing the local traditional culture.

2.2 The Active Development of Public Digital Cultural Service Platform

In order to adapt to the network era of the change of the mode of public access to information, public library at present in China for many provinces are actively explore the "Internet + public culture service mode", through constructing a comprehensive and one-stop public service platform, the digital culture as a whole of the public digital culture resources, expands the scope of service, innovates service mode and improves service efficiency. Such as Jiangsu provincial public digital culture in construction has improved the comprehensive service platform including library system in Jiangsu province library cloud platform, it has integrated and supported the country's public digital cultural projects, public digital culture project in Jiangsu province and the provincial public library resources at all levels and application, for providing one-stop network users in the province [2]. Zhongshan Library of Guangdong Province has designed and developed the "Guangdong Culture E-Station" service platform, which integrates public cultural in formation promotion, mobile digital reading display and dissemination of characteristic cultural resources, and realizes the centralized integration, unified release, and extensive sharing and utilization of public cultural digital resources [3].

3 The Problems Existing in the Construction of Public Library Digital Culture

Although the public libraries in China have made great achievements in digital culture construction, there are still some prominent contradictions and problems, which are mainly summarized as follows:

3.1 Imbalance between Supply and Demand in Resource Construction

In the process of digital resources construction, many public libraries just blindly purchase digital resources and online databases. Although the total amount of public digital resources is very huge, there are some disadvantages to some extent demonstrated in homogenization, unbalanced quality, single type and low ease of use. There are insufficient supply of resources and services that the public is interested in and enjoys. Therefore, the existing resource pool is not attractive enough to the public and cannot fully meet the public's demands for public digital cultural services. In addition, some public libraries only pay attention to serving local readers while constructing digital resources of local culture, ignoring the convenience of non-local readers or researchers to find relevant resources. This kind of traditional service is still limited by region, and not conducive to the full utilization of digital resources with local characteristics.

3.2 The Improvement of Resource Integration

Although all provinces and cities have built regional digital libraries of various scales and a large number of public digital cultural service websites at present, these public digital cultural resources have not been fully combined. Readers want to access digital reading services, sometimes they need to go through the cumbersome retrieval procedures, and cannot really achieve the "one-stop" access. For example, the platform of "Guangdong Culture E-Station" only achieves the aggregation and release of public cultural information resources, but it does not involve the system integration of one-stop management or reservation of public cultural services.

3.3 Lack of Demands Investigation and Feedback Evaluation

Many libraries employ top-down mode rather than demand-oriented mode of resource supply, due that lacking of long-term effective user demands survey and public satisfaction evaluation data to guide, the contents of the digital culture service for providing the public are machine-made, and unable to meet the social public diversification, personalization, multi-level cultural needs, thus it leads to the low utilization of public digital culture resource, and it exists the widespread phenomenon that service efficiency of public digital culture is not high.

3.4 Inadequate Marketing for Existing Services and Insufficient Social Participation

Due that public digital culture service is the product of new era, a lot of public libraries are heavily focused on construction, ignoring service, they lack of experience on how to make use of the magnitude of the digital library culture resources and effectively promote the spread, some libraries only published some dynamic news in their own official website and public Wechat account, the activities form carried out is relatively drab, lacking of innovation and keeping up with the current pace of change in the public demands, therefore it is unable to attract the attentions of the public, leading to low social public participation, furthermore it will be difficult to form large-scale digital service of user groups.

4 The Innovative Roadmap to Public Library Digital Culture Construction

4.1 Promotion on the Supply of Digital Resources

The readers' services for traditional library mainly focus on face-to-face consultation and lending for paper books. In the future, the reader service function of the library will lay more emphasis on the supply of digital resources and the online cultural service mode. Accurately grasping the needs of readers is the fundamental guarantee of the library's continuous prosperity. For example the big success in Japan Daikanyama T-Site, the operator is CCC group (Culture Convenience convenient Club Culture Club) which is actually a big data collection and analysis of the company, they use large reader data to analyze the readership, further recommend relevant books and magazines to readers according to their hobbies, interests and concerns so as to accurately select and sell books. In 2013, Daikanyama T-Site partnered with the Muhsiung City Library brought its service concept to public facilities. In the 13 months since Daikanyama Daikama opened in Takeo, a city with a population of only about 50,000, there had been more than 1 million visitors, making it a prime example of public renovation in Japan [4]. The core point behind its successful operation is to integrate the book-related supply platform with standardized data collection, and respect for individualized research methods and professional content planning. Therefore, the key point of the future library in digital resources construction should be to establish the "user thinking", through big data analysis, the libraries grasp the spiritual and cultural needs of the readers, and provide readers with the "knowledge dry goods".

4.2 The Improvement on Resource Integration

It is improvement on resource integration for the construction of one-stop platform. One-stop platform firstly needs to solve the problem of "viewing" in public digital cultural services, that is, to provide the public with one-stop "aggregation and display of cultural information resources" through technical methods, so that the public can easily see or understand the real valuable cultural resources on the platform [5]. The library cloud platform in Jiangsu province of China receives and grabs the country's public service cloud platform related resources and services data, at the same time, it also uploads characteristic digital resources in Jiangsu province to the cloud directory of China national unified digital resources, so as to complete the public cultural data integration work for local public cultural resources and the service data report to the national center. The second is to solve the problem of "handling" in public digital cultural services. All kinds of public cultural services, such as renewing books, booking lectures, training, exhibition, performance, service evaluation, etc., can be provided by the public in a "one-stop" way on the platform, truly achieving the integration between online and offline public cultural services. For example, Chengdu Public Library Alliance has set up a kind of cluster-type service alliance, which makes the service types and activity forms of 21 public libraries in various districts (cities) and counties of Chengdu to timely update and improve, serving 1.5 million readers a year, and realizing the joint construction and sharing of regional culture [6].

4.3   The Feedback and Evaluation Mechanism

To solve the problem of single and one-sided resource supply mode of public library from top to bottom, online reader feedback and evaluation mechanism should be established by using Internet technology as soon as possible. In recent years, the public cultural service of big data on "collecting like" for order distribution implemented in Shanghai can be regarded as the pioneer of innovative services. Public cultural institutions in Shanghai installed 5,063 QR codes dedicated to neighborhood community exclusive service matching with them. It is put on people's doorsteps, and formed a map of public cultural needs of citizens according to the data of "collecting likes" displayed by the project. The cultural service content selected by the community "order selected" will be accurately delivered to the community gate through the unified content distribution platform [7]. By drawing lessons from this practice to collect readers' feedback and evaluation, users' needs can be easily focused and the service can be precise. At the same time, the "My Library" module is established on the library APP to collect and record the book borrowing history of every reader in different periods with the help of big data technology which is analyzed their reading habits and attention categories, and to push their special services. Readers can also make use of this platform to customize the types of books which they want to read according to their own needs and create a personalized reading environment. At present, the service module of "My Library" has been set up in different types of libraries in Shanghai and Fujian [8]. If such personalized public digital cultural service mode can become more mature and universal, it will greatly improve the user experience of public digital cultural service and promote the overall development of public cultural undertakings.

4.4   "Internet +" Cross-border Library Cooperation

In the past, the traditional form of library publicity to the public was nothing more than the official website, Weibo, WeChat public account, etc., but only with the library's own publicity influence, the amount of public attention is still quite limited. During the Mid-Autumn Festival in 2018, Shanxi Library, together with Shanxi Provincial Committee of the Communist Youth League, Provincial Peking Opera Theatre and Tencent Daqin website, jointly carried out the activity of "Digital Resources of Peking Opera into Campus". The activity was broadcast online by Tencent Daqin website, Tencent Video, China Culture Network TV, "Truth Seeking" Yuandian live broadcast and many other news media. The number of people watching the live broadcast ranked the eighth in the relevant ranking of the National Center for Public Cultural Development of the Ministry of Culture and Tourism of China, and achieved very good social responses. Subsequently, Shanxi Provincial Library and Tencent Daqin.com jointly held a large-scale series of themed activities of "Imagining Shanxi-Folk Welcoming Spring", and carried out stereoscopic propagation of this activity through the new media matrix such as Tencent Daqin.com, Tencent WeChat and Tencent Video. According to the statistics of Tencent background, this activity has been publicized to 18 million Shanxi users, with a total exposure of 17.64 million person-times and nearly 12 million person-times participating in interactive communication, benefiting one-third of the population of Shanxi Province and obtaining the benefits beyond the reach of traditional digital cultural services [9]. Therefore, it can be seen that new media channels of the internet should be actively used to carry out multi-platform cross-border cooperation and give full play to the resource power of all sectors of society, so as to attract more widely attention and participation of the public.

4.5 The Library Brand IP and the Creativity of Service Marketing and Promotion

Library brand IP not only helps to establish a good reputation and dependence among readers, but also facilitates public libraries to carry out cultural service promotion for a long time. Under the background of integration and innovation, the vitality and competitiveness of public library cultural service brand IP come from the people-oriented innovative service. For example, Guangxi Library has successfully created the public digital cultural service brand of "Guangtu Jindouyun". In 2018, it held 17 phases and 26 events in total under the name of the brand, attracting nearly 500,000 people to participate and more than 200 media reports [10]. It has played a good demonstrative role in promoting digital cultural resources in collection and expanding public cultural services. The Greater Victoria Public Library of Canada, which won the IFLA International Marketing Award in 2020, has designed its official homepage into several simple and clear application modules for reading, watching, listening, playing and discovering public digital cultural services. Based on the current outbreak, a novel Coronavirus research database has also been added [11]. The marketing copy of the library, it use a library card to change your mind and change your life, plays a direct impact on the hearts of readers from a higher level of spiritual needs.

5 Conclusion

Under the situation of developing public digital culture vigorously throughout the China, libraries shoulder the important responsibilities of inheriting civilization, spreading culture and serving the society. Only by taking root in the lives of the people, improving the supply system of digital cultural resources, establishing a one-stop resource sharing platform, and providing high-quality, personalized, practical and convenient public digital cultural services, we can meet the growing spiritual and cultural needs of the people and boost the development of China's public cultural undertakings.

Acknowledgements. This research was supported by Research on the Practice of Online and Offline Blended Teaching (No. SZIIT2021SK035), Research on the High Quality Development of Shenzhen Industrial System Under the New Development Pattern of Double Cycle (No. SZIIT2021SK010).

References

1. 'Opinions on Promoting the High-quality Development of Digital Culture Industry' was issued by Ministry of Culture and Tourism. http://www.gov.cn/zhengce/zhengceku/2020-11/27/content_5565316.htm
2. Gengjian: The construction opinion of Provincial public digital cultural service platform: take Jiangsu library cloud platform as a case. Lib. J. Henan **39**(11), 64–66(2019)
3. Wu, H.: Practice and reflection on the one-stop public digital cultural service cloud platform – a case study of the construction of "cultural e-station in guangdong province. J. Lib. Sci, 6 (2019)
4. Masuda, M.: Knowledge of capital: the way Daikanyama Bookstore operates. Culture Convenience Club (2014)
5. Wu, H.: The practice and thinking of new technology application in the construction of public digital culture – also talking about the construction of "public culture cloud." Lib. Sci. Res. Work **1**, 26–30 (2017)
6. Construction of Chengdu Public Library Alliance System in Sustainable Development. Blue Creative Digital Research: http://www.lantrydata.com/a/70450.html
7. Wanyan, D., Wang, Z.: Research on public digital cultural service mode innovation under big data environment. Lib. Inf. **5**, 59–66 (2020)
8. Han, X.: Path analysis of constructing new digital library under the background of public cultural service system. Comp. Stud. Cult. Innov. **4**, 180–181 (2020)
9. Lu, L., Sheng, Q.: Innovative development of public digital cultural services under the background of cultural and tourism integration – a case study of the construction of "intelligent culture cloud landmark" in Shaanxi provincial library. J. Nat. Lib. China **02**, 32–40 (2020)
10. Zhen, L.: A case study of the establishment of public digital cultural service brand Guangtu Jingdouyun. J. Nat. Lib. China **29**, 20–25 (2020)
11. The website of public library of Victoria of Canada. https://www.gvpl.ca/

Author Index

Printed in the United States
by Baker & Taylor Publisher Services

Printed in the United States
by Baker & Taylor Publisher Services